Oceanography - Relationships of the Oceans With the Continents, Their Biodiversity and the Atmosphere

Edited by Leonel Pereira and Miguel A. Pardal

Published in London, United Kingdom

Oceanography - Relationships of the Oceans With the Continents, Their Biodiversity and the Atmosphere
http://dx.doi.org/10.5772/intechopen.104007
Edited by Leonel Pereira and Miguel A. Pardal

Contributors
Aksana M. Parfenova, Alexander Roshdestvensky, Alina Hillinger, Ana Marta Mendes Gonçalves, Andrés Castrillón, Chayonn Marinho, Célia Regina de Gouveia Souza, Diana Pacheco, Elena V. Lasareva, Joseph Harari, João Carlos Marques, João Cotas, Kay Ihle, Leonel Pereira, Madalena Mendes, Matheus Souza Ruiz, Miguel A. Pardal, Miguel Cascais, Pedro Andrade, Regina de Souza Ferreira, Renan Braga Ribeiro, Samuel Hora Yang

First published in London, United Kingdom, 2024 by IntechOpen
IntechOpen is the global imprint of INTECHOPEN LIMITED, registered in England and Wales, registration number: 11086078, 167-169 Great Portland Street, London, W1W 5PF, United Kingdom

British Library Cataloguing-in-Publication Data
A catalogue record for this book is available from the British Library

Additional hard and PDF copies can be obtained from orders@intechopen.com

Oceanography - Relationships of the Oceans With the Continents, Their Biodiversity and the Atmosphere
Edited by Leonel Pereira and Miguel A. Pardal
p. cm.
Print ISBN 978-1-83768-858-6
Online ISBN 978-1-83768-859-3
eBook (PDF) ISBN 978-1-83768-860-9

For EU product safety concerns:
IN TECH d.o.o., Prolaz Marije Krucifikse Kozulić 3, 51000 Rijeka, Croatia,
info@intechopen.com or visit our website at intechopen.com.

Meet the editor

Leonel Pereira has a degree in biology (scientific branch), a Ph.D. in biology (specialty in cellular biology), and a habilitation degree (biosciences, specialty in biotechnology) from the Faculty of Science and Technology of the University of Coimbra, where he is currently an associate professor. In addition to teaching at this university, he is a researcher whose interests focus on the areas of marine biodiversity (algae), marine biotechnology (bioactive compounds from algae), and marine ecology (environmental assessment). Since 2008, he has been the author and editor of the electronic publication MACOI—Portuguese Seaweeds Website (http://www.flordeutopia.pt/macoi/), and since 2018 he has been a member of the editorial board of the scientific journals *Marine Drugs* and *Applied Sciences*, is editor of several special issues of scientific journals (*Journal of Applied Phycology*, *Botanica Marina*, *Marine Drugs*, *Applied Sciences*), was editor or author of more than 20 books, and has authored more than 150 indexed articles and more than 66 chapters of books. He has given more than a hundred lectures and oral communications at various national and international scientific events. He is the coordinator of the master's degree on biodiversity and plant biotechnology, and several national and international research projects. In 1998, he received the Francisco de Holanda award (Honorable Mention) and, more recently, the Rei D. Carlos Sea Award (18th edition), and winner of the CHOICE Award—Outstanding Academic Title Award 2016 for the book *Edible Seaweeds of the World* (CRC Press).

Miguel Ângelo Pardal is a full professor at Universidade de Coimbra, Departamento de Ciências da Vida. He finished his Ph.D. in marine ecology in 1999. His main interests are focused on climate change, environmental degradation, oceanic literacy, marine biodiversity, marine contamination, marine biological resources, fisheries, estuarine integrity, long-term studies, food safety of marine products, and sustainable use of marine resources. Miguel Ângelo Pardal is a co-author of more than 320 articles in SCI journals and co-author of 17 international book chapters. He has supervised 26 Ph.D. students and 60 MSc students and has participated in more than 30 scientific research projects. He has an *h*-factor of 52 and over 9,500 citations on the Web of Science. He participated as a member of several international evaluation research panels in Europe.

Contents

Preface

Welcome to *Oceanography – Relationships of the Oceans with the Continents, Their Biodiversity, and the Atmosphere*. This book delves into the intricate connections between Earth's oceans, continents, biodiversity, and atmosphere, exploring the essential role of oceanography in understanding these relationships.

The study of oceanography is pivotal for comprehending our planet's dynamics, as oceans cover more than 70% of the Earth's surface and influence global climate, biodiversity, and human societies. This science encompasses diverse disciplines, including physical, chemical, biological, and geological aspects of the ocean.

In Chapter 1, we provide an introductory overview of oceanography, highlighting its significance in addressing pressing environmental and societal challenges of the 21st century. From marine currents to coastal dynamics and marine pollution, this chapter sets the stage for a comprehensive exploration of oceanographic themes.

The economic and environmental importance of seaweed aquaculture takes center stage in Chapter 2. We explore how seaweed cultivation contributes to sustainability, the blue economy, and climate adaptation, offering insights into its potential as a tool for conservation amid climate change.

Chapter 3 delves into the deep-sea biodiversity on continents, elucidating how organisms thrive in extreme environments like hydrothermal vents and cold seeps. Recent studies have uncovered fascinating links between tectonic plate dynamics, mineral deposits, and marine life, underscoring the profound impact of oceanic processes on terrestrial ecosystems.

Chapter 4 explores the large-scale interaction of oceans, continents, and the atmosphere in the northern hemisphere. This section investigates seasonal fluctuations in heat and moisture transfer, elucidating the dynamic relationship between these elements on a global scale.

Each chapter is designed to deepen our understanding of oceanographic processes and their broader implications for Earth's systems. From biodiversity to climate adaptation and environmental management, *Oceanography* offers a comprehensive exploration of our planet's interconnectedness.

In Chapter 5, the focus shifts to the river–sea transition zone, investigating the transport of mineral colloids and the influence of organic matter. This chapter sheds light on crucial mechanisms affecting marine ecosystems, emphasizing the role of organic substances in material transport.

Chapter 6 tackles hydrodynamical numerical modeling of coastal areas, examining the complexities of coastal dynamics and their societal implications. The chapter underscores the unique challenges posed by coastal regions and the necessity of effective environmental management. The book concludes with Chapter 6.

We hope this book serves as a valuable resource for students, researchers, and enthusiasts seeking to unravel the mysteries of our oceans and their profound relationships with continents and the atmosphere.

Leonel Pereira and Miguel A. Pardal
Department of Life Sciences,
University of Coimbra,
CFE—Centre for Functional Ecology: Science for People and Planet, Marine Resources, Conservation and Technology,
Coimbra, Portugal

Section 1

Introduction

Chapter 1

Introductory Chapter: General Overview on Oceanography

Leonel Pereira and Miguel A. Pardal

1. Introduction

Spherical, gigantic, bright blue with large whites in areas of clouds, ice, and storm spirals. This is the Earth seen from space by an astronaut. Blue corresponds to the ocean that moderates the temperature, significantly influences the climate, and ensures the maintenance of life as we know it today. The human population has used the ocean since the beginning of the times, and the conquest of the seas is directly related to the evolution of human society, which establishes the movement of people relative to the sea, raw material, food, commodity/trade exchanges, energy, and as a waste disposal area, unfortunately currently contributing to the effects of climate change. About 6,24 billion people (78% of the world's population) live within a radius of 200 km from the sea, making up the largest metropolises on the planet, almost all of which are directly connected to the ocean [1].

Seas and oceans are smaller segments of the only ocean that involves all the emerged land of the planet, divided only for purposes of convenience (social and political) and geolocation. During the Middle Ages, much of the maritime trade was carried out between the Mediterranean sea and other small seas in that region, becoming famous the "Seven Seas," comprising the Adriatic, Arabian, Caspian, Mediterranean, Black and Red sea, and the Persian Gulf [2]. Currently, globalized human civilization, in constant exchange of information, responds with a much broader and more analytical look at the ocean. The new seven seas are made up of the North Pacific, South Pacific, North Atlantic, South Atlantic, Indian, Arctic, and Antarctic oceans. According to the International Hydrographic Organization, there are 61 seas on Earth, such as the Caribbean Sea (Central America), North Sea (Northern Europe), Gulf of Mexico (Mexico and USA), Bering Sea (between America and Asia), Persian Gulf (in the Middle East), and Hudson Bay (Canada and USA) [3].

Seas and oceans together cover 71% of the Earth's surface, which corresponds to an area of 361,100,000 Km^2 and a volume of 1,338,000,000 Km^3 (**Table 1**). Its deepest point is the Mariana Trench, 11,022 m deep in the Pacific, and the highest point is in the Hawaiian Marine Range (USA), a sea mountain 10.203 m high from the ocean floor (**Figure 1**). On average, the ocean has a layer 3796 m thick, with a temperature of 3.9°C and a salinity of 34.482 g of salt per liter of water, usually at a salinity of 35. In comparison, the surface land is only one thickness of 840 m and the Himalayan range (Nepal) with 8848 m. The deepest point, among all continents, is in Siberia (northern Russia), Lake Baikal with 1680 m deep, which dams 20% of the planet's fresh melting water [4].

Portions of water on Earth	**Volume (Km^3)**	**Percentage (%)**
Salt water (Oceans and Seas)	1.338.000.000	97,0
Fresh water	35.000.000	3,0
Portions of water on Earth	Volume (Km^3)	Percentage (%)
Glaciers	24.000.000	68,7
Subterranean water	10.500.000	30,1
Permafrost	300.000	0,8
Surface water and atmosphere	135.000	0,4
Surface water and atmosphere	Volume (Km^3)	Percentage (%)
Lakes	91.000	67,4
Soil moisture	16.500	12,2
Atmosphere	13.000	9,5
Swamp water	11.500	8,5
Rivers	2.120	1,6

The largest portion is known as ocean, being salt water.

Table 1.
The proportions of water mass on the planet are distributed differently throughout the planet, whether in solid (ice), liquid (ocean, rivers, lakes, groundwater), or gaseous (atmospheric vapor) form.

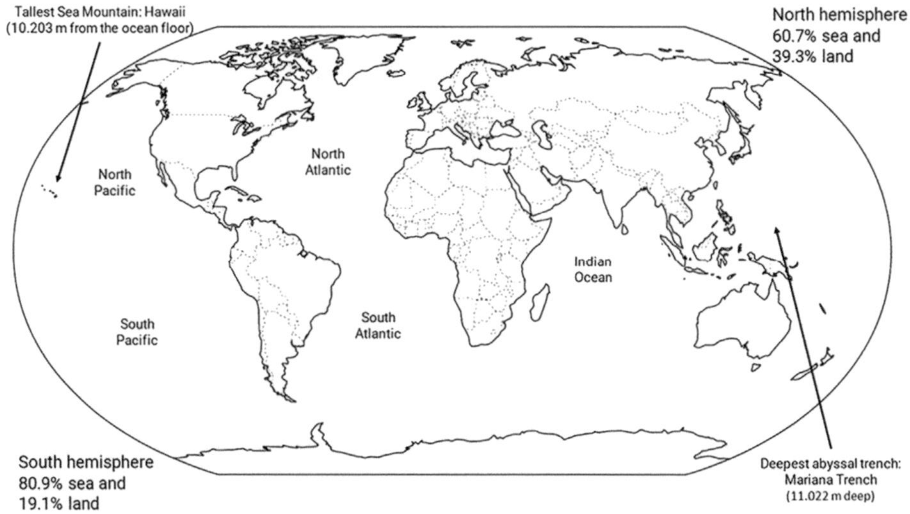

Figure 1.
About 71% of the surface of the planet is covered by the ocean, and this volume corresponds to 97% of all the water in the Earth's crust.

While the ocean may seem incredibly large, on a planetary scale, it is insignificant. In an image that portrays the surface of the Earth covering a paper globe of 12 cm in radius, the oceans would only represent the thin layer of blue ink that colors the paper, considering the 12 cm radius of the planet. The ocean accounts for about 0.02% of the planet's mass. There is an immensely greater volume of water inside the planet than in the ocean, atmosphere, and rivers. The Northern Hemisphere has

60.7% of its surface area of sea and 39.3% of land, the largest portion of emerged land. In the Southern Hemisphere, its largest area is destined to the sea with 80.9%, and only 19.1% of land (**Figure 1**).

Oceanography is important because the oceans play a critical role in global climate, producing oxygen, regulating sea levels, and creating habitats for millions of marine species. In addition, the oceans are a vital source of food, energy, and mineral resources for humanity. For these reasons, oceanography is a fundamental science for understanding and sustainably managing the oceans and their resources [5].

2. First navigations

Historically, civilizations that used maritime transport (mobility or food) had greater development and greater territorial borders in relation to other cultures. The first written records of maritime trade date back to 2000 BC in the Mediterranean sea. The Cretans were the first people to establish maritime supremacy in the Mediterranean. After the fall of this empire in 1200 BC, the Phoenicians gained control and expanded the commercial zone beyond the Straits of Gibraltar. Greek culture began its dominance of the Atlantic Ocean in 900 BC. They were the first to observe a north-south current beyond Gibraltar, considering all this body of water as an immense river, called the "Okeanos." However, these expeditions were closely associated with the coastal zone, very few ventured high sea. On the other side of the world, other peoples also took to the sea, such as the Chinese, who developed a complex waterway system that connected the various rivers to the Pacific Ocean. It is estimated that in 3000 BC. Polynesian peoples already moved easily between the islands of present-day Indonesia and South Asia, starting the colonization of islands in the central portion of the Pacific. These "sailors" were based simply on observing the Sun and stars during dawn and dusk [6].

Trade and the conquest of new lands promoted travel increasingly ambitious, long, and far from the coast. It is undeniable that marine science had its beginnings linked to simple observations described by navigators. In 300 B.C., the largest library in the history of the ancient world was founded, with the largest collection of parchments, the Library of Alexandria (Egypt), considered the first university on the planet. Due to this source of information, marine sciences had a great leap forward in their applied studies. One of the most famous librarians who managed the library was the Greek Eratosthenes of Cyrena, the first to remarkably calculate the circumference of the Earth. Although Pythagoras had already reached the conclusion that the planet was round in 600 BC, it was Eratosthenes who estimated its size. The original figure published by the librarian in 230 B.C. differs by only 8% from the currently calculated real value (40.075 km) [7].

2.1 Scientific navigations

The first documented scientific expedition took place between 1768 and 1771, under the English flag of Captain James Cook, on the ship HMS Endeavor. Although the expedition had several goals, scientific observation was one of them. It was on this expedition that they "discovered" New Zealand and mapped the Great Australian Barrier Reef. The success of his first voyage resulted in two other expeditions: one (between 1772 and 1775) to the extreme south, being the first navigator to circumnavigate the world at high latitudes, which "discovered" Easter Island and reached latitude 71°S, although it did not find Antarctica. His last expedition was between

1776 and 1779, with the aim of exploring the high latitudes of the North (Canada, Alaska, and Siberia). On this voyage, Cook "discovered" Hawaii and charted the west coast of North America. He and scientists from the British Royal Academy collected samples of plants, animals, various marine organisms, and samples from the ocean floor. The detail in the description of your Pacific nautical charts is so accurate that it helped the Allies during World War II [8].

All of the above expeditions promoted great advances in marine science, even though none of them had academic research as their main objective. The first circumnavigation expedition with the exclusive main objective of marine sciences was the British ship HMS Challenger of 1872–1876. Another famous previous expedition, that of the ship HMS Beagle, was commanded by Captain Robert FitzRoy and naturalist Charles Darwin between 1831 and 1836. This resulted in remarkable discoveries for the theory of the evolution of life on the planet, but it was mainly focused on experiments and continental samplings, C. Darwin's bestseller The Origin of Species (published in 1859), was one of the fruits of this splendid expedition, which inspired the future Challenger [9].

3. Physical oceanography

Physical oceanography is the part of oceanography that focuses on understanding the physical properties of the oceans, such as temperature, salinity, current, swell, and depth. These properties affect the circulation of ocean currents, the formation of islands, and continents, marine life, and the interactions of the oceans with the atmosphere and climate [10].

Physical oceanography focuses on understanding how temperature, salinity, and current affect the circulation of ocean currents. Ocean currents are important because they affect the global climate, help transport nutrients, and regulate sea levels. Physical oceanography also studies waves, including how they are generated, how they propagate, and how they affect marine life [11].

Physical oceanography also focuses on understanding the depth of the oceans and undersea topography. This includes the formation of seamounts, island chains, and the cycles of erosion and sediment accumulation. Undersea topography is important because it affects current circulation, carbon dioxide production, and marine life [12].

In addition, physical oceanography studies the interactions between the oceans and the atmosphere, including how the oceans affect global climate and how they are affected by climate. For example, oceans can affect global climate by transporting heat and water vapor, and climate can affect oceans through precipitation, evaporation, and ice formation [13].

In summary, physical oceanography is an important part of oceanography that focuses on understanding the physical properties of the oceans and their interaction with the atmosphere and climate. This is crucial to understanding how the oceans affect and are affected by the global climate and to ensure the sustainable management of marine resources [14].

3.1 Atmospheric circulation

The sun and atmosphere directly or indirectly control almost all dynamic processes within the ocean. The dominant external factors and energy sinks are sunlight, evaporation, infrared radiation emission from the ocean surface, and the ocean's sensible heat from hot and cold winds. Winds control the surface circulation of the ocean down to about 1 km of depth. Wind and tides drive deep ocean currents [15].

The ocean, in turn, is dominated by a heat force that conducts atmospheric circulation differently from the equator to the poles. The uneven distribution of heat balance (loss and gain) across the ocean drives winds through the atmosphere. The sun heats the tropical ocean, which evaporates, transferring heat in the form of water vapor to the atmosphere. Heat is released when the steam condenses into rain. Winds and ocean currents transport heat toward the poles, where it is lost to the atmosphere [16].

Air heats up, expands, and rises at the equator; in the same way that it cools, contracts, and performs downward movement at the poles. However, instead of continuing from the equator to the poles continuously in each hemisphere, air rising at the equator is gradually deflected eastward as it moves poleward, that is. the air turns to the right in the Northern Hemisphere (NH) and to the left in the Southern Hemisphere (SH). This change in direction is caused by the Coriolis effect (a real effect that depends on the frame of reference), which, despite not causing the wind, influences the direction [17].

From the moment the air rises at the equator, a decrease in humidity by precipitation (rain) caused by the cooling and expansion. This drier air then becomes denser in the upper atmosphere as it begins to radiate heat into space and cools. After moving from the equator to about 30°N and 30°S latitude, the air becomes dense enough to descend to the Earth's surface. A large portion of the descending air returns toward the equator when it reaches the surface. In the NH, the Coriolis effect influences the direction of surface air to the right. Despite being warmed by compression during its downward movement, air is normally cooler than the surface it flows through. As a result, the air warms up as it moves toward the equator but evaporates surface water and becomes humid. This moist, heated, and less dense air begins to rise as it approaches the equator, closing the cycle. This significant air loop is called the atmospheric circulation cell. There are two cells in the tropics (0° to 30°): Hadley cells. Two cells at mid-latitudes (between 30° and 50–60°): Ferrel cells and two cells at high latitudes (50–60° up to 90° - poles): polar cells. These three large atmospheric circulation cells described are also represented by the trade winds (northeast and southeast), west winds, and east winds, respectively [18].

This atmospheric circulation model described provides a very interesting understanding for physical oceanography. From understanding of these atmospheric dynamics, it is possible to extend the study to various phenomena or processes that occur in the earth-ocean-atmosphere system, such as monsoons, breezes (sea and land), storms, cyclones (tropical and extratropical), and even phenomena, such as "El Niño" and "La Niña" (also correlated with oceanic circulation) [19].

3.2 Oceanic circulation

As seen in the description of atmospheric circulation, there is an energy or heat balance between the equator and the poles through the atmosphere and oceans. This interface is extremely important. This energetic (or thermal) balance is essential for the dynamics of winds and ocean circulation. Energy transport across the oceans *via* ocean currents accounts for 10 to 20% of the heat distribution across the planet. Basically, sea water moves in currents, either shallow or deep. Surface currents affect only the shallowest tenth of the oceans, and their movement is influenced by heat balance and winds. In general, the movement of surface currents is horizontal, and they can also flow vertically according to the wind that blows near coastal regions or along the equatorial region. Surface currents flowing from the equator (low latitudes) transport heat to the poles (high latitudes), nutrients, and influence climate and weather. In addition, they are essential for navigation. Deep oceanic or thermohaline circulation is driven by density differences between water masses. Remembering that the density in the oceans is defined by the relationship between temperature, salinity, and pressure (due to the great depths). This circulation accounts for 90% of sea water below the surface layer [20].

In general, the Coriolis effect, the force of gravity and friction influence movement (direction, up and down, and intensity/velocity) of surface and deep ocean currents (thermohaline). The oceans are interconnected but do not perform significant water exchange between them, and this fact occurs because the water masses have different oceanographic characteristics (temperature, conductivity, salinity, heat balance), wave dynamics, tides, and currents that differ throughout the planet. Therefore, the oceans are divided into five large portions: Atlantic, Pacific, Indian, Arctic, and Antarctic [21].

4. Geological oceanography

Geological oceanography is an interdisciplinary area that combines knowledge from geology, oceanography, geophysics, geochemistry, geomorphology, and sedimentology. Its aim is to understand how the geological features of the oceans and seafloor emerged and how they evolved over time [22].

One of the main techniques used in geological oceanography is seismic, which involves the emission of sound waves that are reflected from different underground layers, allowing scientists to obtain a three-dimensional image of the marine subsoil. Other techniques include taking sediment samples, using sonar to map the seafloor, and conducting underwater drilling [23].

Geological oceanography is also important for understanding the evolution of undersea sedimentary basins, which are large areas of sediment deposition, and for studying undersea volcanic eruptions and undersea earthquakes. This information is critical for predicting potentially dangerous geological events, such as tsunamis, and for properly managing coastal zones and marine resources, such as oil and natural gas [24].

The coastline is one of the most dynamic natural features on the planet. Its position in space constantly changes on temporal scales of seconds (waves), hourly (high tides and low tides), daily (storms), seasonal (seasons of the year), annual (El Niño), decadal, secular, and millennial. The daily rise and fall of sea level and other bodies of water connected to the ocean (estuaries, lagoons, etc.) are caused by the interference of the Moon and the Sun on the Earth's gravitational field. The amplitude of the tides (the difference in level between high and low tide) is a modeling element of the coastline, as a function of the current velocities associated with it. These tidal currents are significant in coastal sediment transport. Tidal currents have the capacity to modify the morphology of the coastline and the inner continental shelf [25].

Geological oceanography also plays an important role in understanding marine biodiversity and protecting marine ecosystems. Undersea geology can influence ocean currents, temperature, and the chemical composition of water, which has implications for marine life. In addition, geological oceanography can help identify and protect areas with high biodiversity and critical habitats for marine life [26].

In summary, geological oceanography is an important field that provides a fundamental understanding of the evolution of the Earth and oceans, as well as the interactions between human activities and the marine environment [14].

5. Chemical oceanography

Chemical oceanography is the branch of oceanography that focuses on the study of the chemical composition of the oceans and their relationship to the Earth and climate. The oceans are an important source of elements and chemical compounds that are essential for life on Earth, and chemical oceanography seeks to understand how these elements and compounds are distributed, transported, and transformed within the oceans [27].

Chemical oceanography focuses on issues such as the variation in the concentration of salts, the formation of chlorides, the concentration of carbon and other elements, as well as the interaction of these components with the atmosphere, continents, and marine life. In addition, chemical oceanography also investigates the chemistry of the deep oceans, including the chemical composition of the water, the concentration of metals, and the presence of organic compounds [28].

5.1 Water

Water is a molecule formed by chemical bonds between two hydrogen atoms and one oxygen. Because of the arrangement of hydrogen and oxygen atoms, water has angular geometry, making it polar. Thus, its positive part attracts negative particles, while the negative attracts positive particles. Water has intriguing characteristics, which make it a compound, despite being abundant, so special and unique [29]:

- High specific heat: Water needs an amount of high energy to raise the temperature of 1 gram of substance by one degree Celsius (°C). Therefore, water resists temperature changes and works as a climate moderator for the planet, transporting heat from low-latitude regions to high latitudes.

- High latent heat of fusion: It is a large amount of energy gained or lost for water in the solid phase to pass into the liquid phase, or vice versa, without having an increase in temperature.

- High latent heat of vaporization: Tt is large amount of energy acquired or lost so that the water in the liquid phase passes to the gaseous phase, or vice versa, without having an increase in temperature.

- Density: The density of pure water is 1000 g/cm^3; already, the water sea has a density between 1.020 and 1.030 g/cm^3. Water density is strongly related to salinity and temperature. Therefore, the density of sea water increases with increasing salinity and pressure and with decreasing temperature. Thus, the colder and saline water is, the denser it is, and the warmer and less saline it is, the less dense it is.

- Sound: The speed of sound waves in sea water is approximately 1.500 m per second, much higher than the speed of sound in air, which is why many marine animals use sound to orient themselves in the oceans.

- Light: The water layer can be divided into two denominations relating to lighting: photic zone and aphotic zone. The photic zone is the layer illuminated and heated by the Sun on the surface of the water, allowing all photosynthetic production. The waters located below this zone do not receive illumination, which is why it is called the aphotic zone.

- Effective solvent: The previously described electrical asymmetry of the water molecule explains its great solvent capacity, being able to attract negative or positive particles. Water is the substance capable of dissolving more substances than any other.

- Salinity: It is the total amount of inorganic solids dissolved in the water. The average salinity of the oceans is 3.5%, depending on evaporation, precipitation, and the amount of fresh water discharged from the continent. Sodium chloride and sulfate are the most abundant dissolved ions in sea water. The determination of salinity is done through the chlorinity of the water sample. Since the ratio of chlorinity and salinity are constant, the total mass of bromide, iodide, and chloride ions is measured.

- Composition of sea water: Sea water is added water solids and dissolved gases, which can be classified as conservative elements, trace elements, nutrients, and dissolved gases. Conservative elements are those that appear in high concentrations. Trace elements are those that, although found in very low concentrations, play an important role in water chemistry and biota. The nutrients are very important for the biota, mainly for the organisms at the bottom of the food chain, affecting all subsequent trophic levels. They generally have low values in surface waters due to the rapid consumption of photosynthetic agents, and in deeper waters, there is an accumulation of nutrients. The gases present in the air dissolve easily in water; thus, all gases are dissolved at the surface of sea water. Among the dissolved gases, the most representative are nitrogen, oxygen, and carbon dioxide.

- pH: In a volume of water, the more hydron (H^+) and the less hydroxyls (OH^-) the more acidic it will be; therefore, the reverse makes it more alkaline. The pH scale ranges from 1 to 14; the value 7 indicates the neutral point at which there is balance in the amount of H^+ and OH^- ions, while the predominance of H^+ ions indicates low pH values; the opposite indicates high values. Pure water is neutral, whereas sea water is slightly alkaline, ranging from 7.4 to 8.5. This is due to the large amount of CO_2 dissolved in sea water, which can produce H^+, bicarbonate (HCO_3^-), and carbonate ($CO3_2^-$) ions, and which prevent pH change when acids or bases are added. Thus, sea water acts as a buffer solution.

Marine systems undergo changes in their chemical composition due to changes in the environment, usually associated with the entry of contaminants. It should be remembered that the environmental quality can be altered by the presence of these toxic agents. The main anthropogenic inputs found refer to the dumping of dredged material, urban and industrial effluents, leaching from rural areas, atmospheric inputs, and shrimp farming waste [30].

5.2 Chemical oceanography and the climate change

Chemical oceanography has important implications for understanding climate as the oceans are an important climate regulator, storing and transporting large amounts of heat and carbon dioxide. Chemical oceanography is also important for understanding biogeochemical cycles, including the carbon cycle and nutrient cycle, which are fundamental to marine life [31].

Chemical oceanographers use a variety of techniques to collect and analyze samples of water and marine sediments, including chemical analysis techniques, spectroscopy, and remote sensing. This information is important for the management of marine resources such as fisheries, tourism, and oil and natural gas exploration, as well as for understanding marine biodiversity and the health of marine ecosystems [32].

In summary, chemical oceanography is an important field that provides a fundamental understanding of ocean chemistry and its relationship to climate, marine life, and the wider environment.

6. Biological oceanography

Biological oceanography encompasses a wide range of disciplines from molecular biology to ecology to better understand the functioning of marine ecosystems and their inhabitants. Some key areas of study include [33]:

- Marine biodiversity: The study of the diversity of species and their distribution in the ocean.
- Marine ecology: The study of how marine organisms interact with each other and their environment.
- Marine phytoplankton: The study of microscopic plants that float in the ocean and form the base of the marine food chain.
- Marine zooplankton: The study of microscopic animals that float in the ocean and are an important food source for many larger marine animals.
- Marine primary productivity: The study of how marine plants produce organic matter through photosynthesis, which supports the entire marine food chain.
- Marine food webs: The study of the complex network of relationships between predators and their prey in the marine environment.
- Marine biogeochemical cycles: The study of the movement of important elements and compounds, such as carbon, nitrogen, and phosphorus, through the marine environment and the impact on marine life.

Biological oceanography also investigates the impact of human activities on the marine environment and its inhabitants, such as the effects of pollution, overfishing, and climate change [34].

The energy that most marine organisms need to survive comes directly or indirectly from the Sun. This produces enormous amounts of energy, which is captured by the chlorophyll present in organisms called primary producers. Solar energy is then transformed into chemical energy. From these reactions, energy is used to synthesize carbohydrates and other organic molecules that will be used by the producers themselves or will be ingested by other microorganisms present in the aquatic environment called consumers [35].

Photosynthesis is the main autotrophic process carried out by beings possessing chlorophyll, represented by plants, some protists, photosynthetic bacteria, and cyanobacteria. The rate of photosynthesis varies depending on the available light (intensity and quality), in addition to other factors, such as the amount of biomass, nutrients, etc. This process is considered dominant in the conversion of energy into carbohydrates, but energy production may also occur through inorganic molecules available in the medium instead of sunlight. We call this process chemosynthesis, and it is present in some marine forms, but it is small in relation to photosynthesis [36, 37].

Primary productivity, synthesis of organic matter from inorganic substances, is expressed in grams of carbon assimilated into organic matter per square meter of ocean surface per year ($gC/m^2/year$). In this context, we call primary producers the autotrophic organisms capable of synthesizing food. The heterotrophic animals that consume these organisms are called secondary consumers, while the animals that feed on these organisms are called tertiary consumers, and so on up to the top consumers. It is important to note that as energy flows, much of it is lost as heat. The energetic interactions between producers and consumers are generally complex, which is why they are called food webs [38].

6.1 Pelagic organisms

Pelagic organisms live suspended in sea water, adrift, and interact in this place with members of very different sizes and characteristics. In this habitat, they share the need to maintain an upright position, in addition to obtaining food, among other basic needs. According to the form of life, we can divide the pelagic organisms into plankton and nekton [39].

Plankton is constituted by the group of beings that live suspended in the water, carried by the current, in aquatic environments of fresh, brackish, and marine water. Many of these organisms have their own movement but insufficient to overcome the force of the currents. Plankton organisms range from micrometers invisible to our eyes without the use of a microscope, to millimeters, with some exceptions of organisms that can reach from several centimeters to meters in length, such as the "Portuguese Man-of-War" (Physalia physalis) [40].

Plankton can be grouped depending on some characteristics:

- Basic cell structure: Prokaryotes or eukaryotes.

- Trophic grade: Autotrophic or heterotrophic.

- Nutrition: Photoautotrophic (light capture and carbon absorption inorganic), osmo-trophy (absorption of organic molecules), phagocytosis or phago-trophy (ingestion of particles)

- Horizontal distribution: Patches or aggregates.

- Planktonic microorganisms: Bacterioplankton (prokaryotes heterotrophs and autotrophs), phytoplankton (eukaryotic organisms, unicellular autotrophs), and proto-zooplankton (eukaryotic organisms, unicellular heterotrophs, Protozoa), zooplankton (heterotrophic eukaryotes organisms).

The phytoplankton community guarantees the ecological balance of the aquatic environment as it forms the basis of the trophic network and of all biological production in the seas. It is a group with a high degree of biodiversity, and new species are isolated and described every day. The microbial food chain of marine ecosystems extends throughout the entire photic zone of the oceans (where the presence of light reaches, up to 200 m deep on average), and it is also known as the microbial loop [41].

The trophic relationships of the loop group different microorganisms, which are mainly responsible for the processes of decomposition and remineralization of the compounds of the biogeochemical cycles [42].

Several other ecological actions can be performed by phytoplankton (adapted from [43]):

- Primary production in the oceans: This is directly related with maintenance of gas balance (CO_2-O_2). Phytoplankton, due to its ability to use CO_2 and release O_2, acts as an important factor in maintaining the balance in the concentration of these gases in water, even influencing physical-chemical characteristics, such as pH.

- Sedimentation of the material produced (biological pump): Through the sedimentation of organic matter to the bottom of the sea (through the death of microorganisms and the release of feces, mainly zooplankton), their removal from the pelagic environment for a long time occurs.

- Carbon removal is important as a climate regulator. In addition, the sedimentation of certain organisms gives rise to the formation of siliceous mud or diatomite (diatoms, radiolarians) and calcareous oozes (foraminifera). Silica from diatomaceous Earth is a chemically inert material and is used as an abrasive substance in filtration in sugar refineries, breweries, winemaking, base for dyes, and insulating material.

- Harmful phytoplankton blooms (HAB): They occur due to a large increase in cell concentration, with reduction of O_2 (at night, or in senescence), or by the production of toxins that are accumulated along the trophic chain.

Marine biodiversity is high in the pelagic environment as the plankton has a high species richness in addition to being in large numbers, usually greater than 30 coinhabitants in space and time unlike terrestrial ecosystems. This ability to inhabit the same ecological niche seems to be a paradox, but it can be explained by the great heterogeneity in the pelagic environment, providing the existence of micro-niches for different species [44].

The lack of knowledge about the life cycle of many planktonic microorganisms makes it difficult to recognize their real species. Ideally, organisms are identified based on morphological criteria (shape and size), ultrastructure, biochemical makeup (chemotaxonomy), and genetics. There is a consensus that the genetic constitution of organisms, details of ultrastructure, and the life cycle should, together, serve as a reference to determine phylogeny, and that phylogeny should determine taxonomy. Thus, the nomenclature should reflect the genome information. This ideal is still far away for protist and prokaryotic organisms and, in practice, the morphospecies are still of great importance in determining the diversity of phytoplankton and proto-zooplankton [45].

How many species are there? In addition to most microscopic species still being unknown, the exact number is unknown, as new species are described daily. Major advances were determined by the introduction of electron microscopy around 1970 and currently by molecular biology. There are at least eight major groups of phytoplanktonic organisms, the most important of which are diatoms (Bacillariophyta) and dinoflagellates (Miozoa). Other prominent groups are Coccolithophyceae and Cyanobacteria [46].

Primary production (PP) in the oceans can be divided into primary gross (the total organic matter produced, excluding the cellular respiration that occurred during the given period) and net primary (organic matter produced is accounted for discounting the "loss" by respiration). Some physical factors may interfere with primary production, such as the primary production rate approximately following the behavior of light with increasing depth of the environment (negative exponential curve). In the euphotic layer, the rate of photosynthesis is high, decreasing to the boundary of the euphotic layer, where the rate of photosynthesis approximately corresponds to the rate of respiration (net primary production = zero). Below the euphotic layer, respiration exceeds photosynthesis and therefore autotrophic cells do not grow [47].

In the marine environment, water column stability is important in phytoplankton and macroalgae (in coastal waters) ecology. The presence of a thermocline or halocline determines stability in the surface and illuminated layer, allowing phytoplankton cells to be exposed to light and providing high gross and net primary production. In conditions of vertical mixing in the water column (isotherms and isohalines), phytoplankton cells are displaced, remaining part of the time in the dysphotic (low light) or even aphotic layer, thus decreasing primary production. The relationship between light, primary production, and respiration in the water column is described by the Sverdrup model. In this model, the concept of critical depth (first defined by Sverdrup, 1953) stands out, that is. the depth at which the production of the entire water column is equal to the total respiration (**Figure 2**) [48].

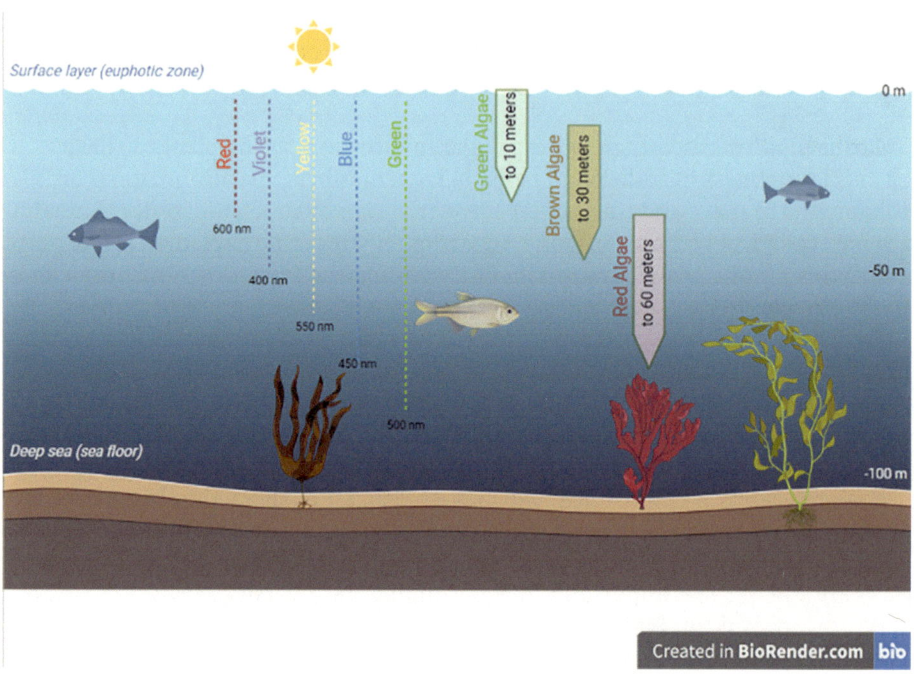

Figure 2.
Light penetration (of different wavelengths) in the sea water column (coastal zone) and the vertical distribution of macroalgae.

Chemical factors also influence PP since phytoplankton actively incorporate inorganic nutrients by using enzymes associated with the cell membrane. The main nutrients or elements required for nutrition can be classified in different ways [49]:

- Macroelements: They are found in relatively low concentrations large, being rarely limiting in the environment: CO_2, Na^+, K^+, Mg^{2+}, Ca^{2+}, and SO_4^{2-}.
- Micronutrients: They are found in lower concentrations, often limiting in the medium: NO_3, NO_2, NH_4, PO_4, and $Si(OH)_4$.
- Trace elements: They are present in minimal amounts, and may be limiting in some cases: Fe, Mn, Zn, Cu, and Co.
- Organic compounds (required by some organisms): Vitamins (B_{12}, thiamine, and biotin), organic binders, urea, and amino acids.

Another group of great ecological importance in plankton is the called zooplankton, in which heterotrophic organisms that feed on primary producers and other zooplankton organisms participate. Formed by animals and larvae of numerous species, the vast majority microscopic, have a certain ability to move in the oceans and seas. The locomotion capacity of zooplankton can be verified with the vertical migrations present in some organisms. It can be classified into two groups:

- Holoplankton: Organisms that spend their entire life cycle in the plankton, such as copepods (the most diverse class of crustaceans, and the largest group that make up the zooplankton), among other crustaceans (such as krill), filter feeder urochordates (salps), chaetognaths, and hydromedusas.
- Meroplankton: Organisms that go through only one stage of life in plankton, such as eggs, larvae, and animals in the juvenile stage, such as larvae of crustaceans, mollusks, and echinoderms. The larvae and eggs of fish are part of the meroplankton and are called ichthyoplankton.

Most of the organisms that make up the zooplankton feed on microalgae, although carnivores, omnivores, and detritivores are observed in addition to herbivores. On the other hand, they are food for many species of fish and other animals, such as the whale, which feeds almost exclusively on krill (pelagic arthropod), considered a key species in the Antarctic ecosystem. They are considered essential organisms for the maintenance of the aquatic ecosystem as are at the base of the food chain [50].

6.2 Nektonic organisms

Pelagic animals, which actively swim in the water column, are known as nektonic beings. Most are vertebrates (mainly fish), but some invertebrates are present in this classification, such as squid and some crustaceans. Turtles and mammals can be important species in certain areas, as can seabirds, especially as predators. The main groups of nektonic predators are:

- Cephalopods: They are the most evolved animals among mollusks, with many marine predators formed by squid, *Nautilus,* and *Octopus*. They are a source of great fisheries.

- Marine reptiles: Turtles, sea snakes, sea lizards (iguanas), and sea crocodile. The best known and most successful are the eight species of sea turtles, possibly due to the presence of their shell, which provides efficient defense. Although the only predator is man, there is attention to the spawning stages, in which there is a high mortality rate and the young when they are on their way to the sea.

- Seabirds: Of four groups of seabirds, the gull and the pelican are the best known as there are many species and these live close to the coast. However, the groups best adapted to the pelagic world are the albatrosses, petrels (Tubinares order, considered the most oceanic birds in the world), and penguins. Penguins, on the other hand, have lost the ability to fly; however, they use their wings to swim over long distances and with great dexterity of movement. Native to the southern hemisphere, they have adaptations to conserve heat, such as insulation with fabric adipose and fat feathers. Another important point to highlight is the presence of large colonies in areas of high productivity, such as the resurgence in Peru. Unfortunately, they are directly affected, with high mortalities, in pollution events (e.g., pesticides) and oil spills. They prey on squid, fish, and zooplankton.

- Fish: They are the group that dominates the nekton, characterized by great ability in swimming, through which they can move independent of ocean currents. There are more species of fish, and more individuals, than there are species and individuals of all other vertebrate groups combined. Some species have modifications that allow them to detect prey or avoid predators. They are divided into two large groups based on the material that makes up their skeletons: cartilaginous and bony. The class Chondrichthyes includes sharks, lampreys, rays, and and is characterized by cartilaginous tissue, jaws with teeth, pairs of fins, and an active lifestyle. Through vibrations in the water, detected by sensitive organs lined up below the surface of the skin, the tiger shark, and the hammerhead shark, among others, attract their prey. In addition, they have a very developed sense of smell, acting in hunting prey, which is composed of fish and marine mammals. In the class, Osteichthyes are approximately 99% of all fish today. In the Teleostei order are the organisms of greatest commercial interest: cod, tuna, and sole, among others. These animals feed on different types of prey depending on their size, location, and prey availability and may be planktivorous, piscivores, or both. They can perform migrations after prey. The vast majority are found in temperate waters, but species diversity is greatest in tropical and subtropical regions.

- Marine mammals: The Mammalia class is the most evolved group of vertebrates, including seals, whales, dolphins, and walruses, and constitutes a very diverse group of 128 species that depend on the oceans for their existence. This group does not correspond to a distinct biological group, but to a functional group that has in common dependence on the aquatic environment for food. This dependency manifests itself on several levels, not all being expressed by the totality of species. For example, dolphins and whales (cetaceans) are completely dependent on the marine environment at all stages of their lives, whereas seals feed in the ocean but breed on land. Marine mammals are divided into four groups: cetaceans, pinnipeds (seals, sea lions, and walruses), sirenians (manatees and dugongs), and fissipeds (carnivores with spread toes, such as the polar bear (*Ursus maritimus*) and two species of otter. They play a fundamental role in the maintenance and regulation of marine ecosystems, especially by regulating the populations of their prey species. These two factors, relevant global biomass and regulatory role, make them a fundamental component of the marine environment. This fact makes them if particularly important if we consider that, currently, about 23% of marine mammal species are threatened [51].

All marine mammals share four common characteristics: a hydrodynamic body, with appendages adapted for swimming; modified breathing system to retain large amounts of oxygen during long dives and displacements. Another important point is the ability to generate internal heat through high metabolic rates and conserve it through insulating layers of fat, and in some cases, even with the presence of hair. The fourth adaptation is related to the absence of need for fresh water due to the ability of their kidneys to excrete urine concentrated in salts, allowing the water necessary for their metabolism to derive from the oxidation of food [52].

The order Cetacea (cetaceans and sirenians) has the only mammals' marine animals that spend their entire lives in the water. Unlike pinnipeds, which mainly use hair as thermal insulation, cetaceans have a thick layer of fat, the "blubber." Hind limbs are absent, and propulsion is provided by horizontal caudal fins. The forelimbs do not have externally individualized fingers, having the shape of oars, and are used to maintain stability during swimming. They are divided into two suborders: Odontoceti and Mysticeti. Current true whales (raw whales) are characterized by their highly differentiated feeding apparatus due to the loss of teeth and the appearance of cornified epithelial tissue plates (fins) that are suspended from the roof of the mouth and serve to filter food from the water. This suborder Mysticeti includes right whales, gray whales, and pygmy right whales, while the suborder Odontoceti includes beaked whales, sperm whales, porpoises, Amazon River dolphins, killer whales, pilot whales, and belugas, as well as of dolphins [53].

6.3 Benthic organisms

They are those animals and plants that live associated with the sediment. Some can burrow (infauna), and others live on the surface of the sediment (epifauna). The benthic habitat can be shallow or deep, full of food, or somewhat barren; the fact is that the diversity of benthic habitats, and of organisms that live associated with them, is very large. Macroalgae forests, rocky intertidal zones, sandy beaches, marshes, and even coral reefs are part of this vast habitat [54].

The epifauna comprises the animals that live on or associated with rocks, stones, shells, vegetation, or on unconsolidated backgrounds. The infauna comprises all animals that live within the unconsolidated substrate layer, drilling into it, or simply living within it. There are still other ways to classify benthic organisms:

- Macrofauna: Comprises all animals that are retained in a 0.5 mm mesh sieve.

- Mesofauna: Animals that pass through a 0.5 mm mesh sieve but are retained in a 0.05 mm mesh.

- Microfauna- all other organisms (generally Protozoa): To survive in the benthic habitat, there are several functional adaptations, such as tolerance to different physicochemical aspects of different types of substrates, as well as structural adaptations, such as body shape and small and elongated bodies (vermiform) are common. The metabolism and activity can be modified by the substrate *via* nutritional aspects (ex: Filters are more frequent in sandy bottoms, where their filtration devices are not at risk of being clogged). In this way, the type of substrate, where the animal lives can modify the rate and forms of reproduction. The horizontal distribution on sandy bottoms is affected by the nature and size of the grains; the type, quantity, and form of organic matter associated with the substrate; the total area of the sandy substrate; and other environmental factors such as water movement, light, salinity, oxygen supply, and pressure.

7. Bioactive marine natural products

One of the fields with a successful history within Marine biotechnology is the bioprospecting of bioactive marine natural products. This subarea stands out as one of the most developed, with several substances being used commercially, and many others in preclinical and experimental testing phases. Until the 1950s, the marine environment went unnoticed by natural product scientists, mainly due to its difficult access. But it was from the 1970s onwards, with the advancement of diving techniques and equipment, that marine organisms became part of chemistry and pharmacology laboratories, starting their history [55].

The first efforts in the exploration of marine natural products were focused on easily available and collectible organisms, such as brown, green, and red algae, sponges, and soft corals, which quickly showed to produce a great variety of new molecules. With continued exploration, in partnership with the progress of oceanographic technology, other groups of organisms, more critical in terms of availability, were studied, and the arsenal of unique molecules from the marine environment grew [56].

In general, the exploration of marine natural products resulted and continues to result in the discovery of many substances, most designated for the pharmaceutical industry, but also occupied space in the areas of cosmetics, agriculture (pesticides), and shipbuilding (antifouling's) [57].

7.1 Biopolimers

Products made from nonbiodegradable polymeric materials (such as plastic), which come from fossil sources, have become a problem due to the growing number of inappropriate discards, and the degradation time of these materials, which take many years in the environment. Researchers have been looking for alternatives, together with the industry, to minimize the environmental impacts caused by the inappropriate disposal of plastic products. Among the alternatives, in addition to reuse and recycling, the production and use of biopolymers, biodegradable polymers, and green polymers have been growing due to their technical and economic viability, presenting great potential for expansion. They can come from renewable sources, such as cellulose, potatoes, or be synthesized by bacteria from small molecules, or even be derived from animal sources, such as chitin or proteins [58].

Crustaceans, such as shrimp and crabs, produce chitin, the second most abundant polysaccharide in nature, after cellulose. Chitin is a very versatile substance for industrial application, in addition to being useful in other areas. It is used in the composition of agricultural fungicides. In medicine, it is used for the manufacture of hemodialysis membranes, in biodegradable surgical threads, as substitutes for artificial skin, healing for burns, and medicine capsules and insulin releasers. In cosmetics, chitin is used in the manufacture of shaving creams and moisturizing creams. Due to its ability to absorb fats, chitin is present in the composition of several dietary foods. It is also used in papermaking and the textile industry. The flocculant and coagulant actions of chitin are applied in the filtration of water in swimming pools, in water sanitation, and in the removal of heavy metals and oils [59].

Marine algae (seaweeds) represent one of the most abundant sources of relevant and widely used biopolymers. Agar, used in research as a raw material for gels and biological matrices; carrageenans, used as stabilizers and texturizers by the food industry and in cosmetic and hygiene product formulations; alginic acid, used as a biomaterial in medical sciences for skin grafts, dressings, and healing agents for serious cases, such as burns, as well as a vehicle for administering drugs or gene therapy and as a base in the preparation of dishes in gastronomy (**Figure 3**). Some mussels and barnacles have been explored for their adhesive properties for fixing on consolidated substrates. This type of "glue," produced mainly by barnacles, has been used in surgical procedures, replacing the suture [60].

Seaweeds are used in many countries for very different purposes: directly as food, extraction of phycocolloids, extraction of compounds with antiviral, antibacterial or antitumor activity, and as biofertilizers. Seaweed polysaccharides (phycocolloids), such as agar, alginates, and carrageenans, are produced on a large scale and have a wide range of applications in the food, pharmaceutical, and cosmetic industries.

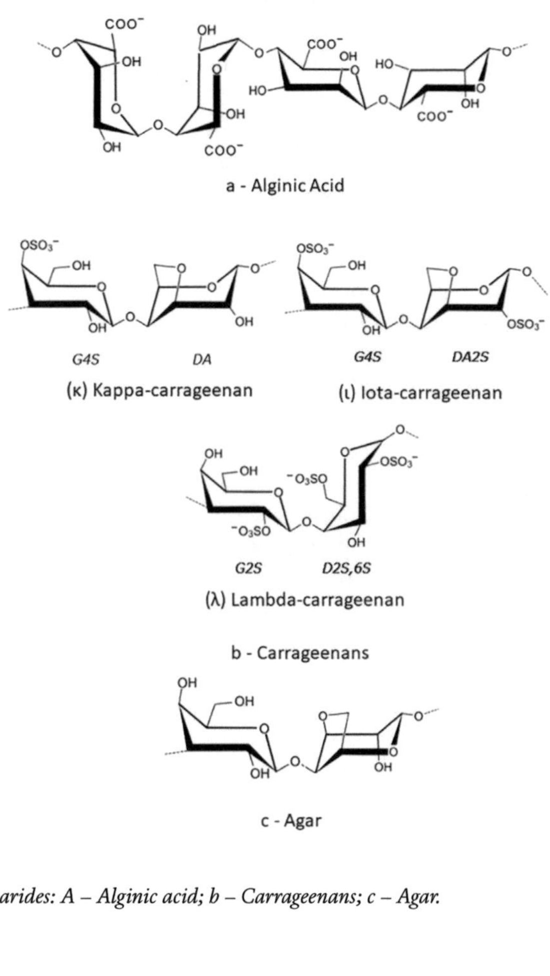

Figure 3.
Main seaweed polysaccharides: A – Alginic acid; b – Carrageenans; c – Agar.

Many species of seaweed (macroalgae) are used as food and have also found use in traditional medicine because of their perceived health benefits. Seaweed is a rich source of sulfated polysaccharides, including some that have become valuable additives in the food industry because of their rheological properties as gelling and thickening agents (e.g., alginates, agar-agar, and carrageenan). The different phycocolloids used in the food industry as natural additives are (European phycocolloid codes): alginic acid, sodium alginate, potassium alginate, ammonium alginate, calcium alginate, propylene glycol alginate, agar, carrageenan, semi-refined carrageenan or processed *Eucheuma* species, and furcelleran [61, 62].

7.2 Biofuels

Fossil fuels are responsible for the emission of gases that intensify the greenhouse effect (heating of the earth's atmosphere). The gravity of this fact could be minimized through the indirect use of solar energy to obtain fuels derived from photosynthetic organisms, which can be cultivated practically all over the world, in a renewable and nonpolluting way. Recent research indicates that the production of biodiesel from microalgae could radically change the fuel market. With a much higher oil production potential per area than traditional crops grown on land, microalgae have aroused worldwide interest in prospecting for biofuels [63].

The advantages arising from biodiesel from algae include their fast growth rates and their high yield per hectare, the fact that they do not contain sulfur, are nontoxic, and are highly biodegradable. Productivity is higher in controlled environments (photobioreactors), but other forms of production are also superior for open systems. Significant investment in research is still needed before high levels of productivity can be guaranteed on a commercial scale. In addition to producing oils, algae are also rich sources of vitamins, proteins, and carbohydrates. Several companies and universities are involved in algae biofuels, and in 2020, the development of a systemic approach to the sustainable commercialization of this biodiesel and its bioproducts was announced in USA [64].

In addition to biodiesel, bioethanol also paves the way within the marine biofuels, being produced from cellulose and alginate extracted from macroalgae or from tunicin, a component present in the tunic of ascidians. Sea squirts also contribute to methane production from their biomass [65].

7.3 Bioremediation

Ecological systems have a level of innate ability to break down contaminants or pollutants that adhere to them. The biological agents responsible for these automatic cleanings are often microorganisms from nature itself. The elimination or breakdown of environmental contaminants by living organisms is called bioremediation. Such microbial-mediated removal over time can take place completely without human intervention; however, the process can also be initiated by anthropogenic administration. The greatest bioremediation efforts in the marine environment are focused on oil spills or other petroleum product contamination, where Marine biotechnology can play a significant role in the final stages of total cleanup [66].

For this case, there are three bioremediation strategies used:

- Intrinsic bioremediation is the removal of oil naturally by biotic means from the environment itself over time and without human intervention.

- Bio-stimulation, which needs native microbial populations to degrade contamination. It is carried out by human intervention through the addition of nutrient fertilizer or other means that increase the rate of natural biodegradation.

- Bio-augmentation, which is the least common strategy and consists of adding an oil-degrading microbiota that complements the degradation capacity of native populations. It may be a rare lineage or one absent from the local community, or even genetically created to carry out the function of degradation. It is not widely practiced due to the environmental concerns of such an intervention [67].

One of the biggest concerns is the toxic polyaromatic hydrocarbons (PAHs) that make up the tar, found in the oil. Through DNA fingerprinting techniques, researchers have isolated marine bacteria that degrade PAHs. Currently, efforts are being made to understand how communities of natural bacteria can detoxify areas contaminated by hydrocarbons, as well as unravel microbial metabolism and growth in contaminated environments [68].

8. Conclusions

This chapter covers the main aspects of oceanography, the interdisciplinary science that studies the oceans, including their biology, geology, physics, and chemistry. It is an important area for understanding climate change, preserving marine resources, predicting natural events, such as tsunamis, and for safe navigation and fishing. Oceanography is also crucial to understanding Earth's dynamics, and how the oceans affect the global climate. However, much remains to be discovered about the oceans, and oceanography remains an ever-evolving field with frequent discoveries and technological advances. In short, oceanography is a vital area for understanding and protecting our planet and its marine life.

Acknowledgements

Leonel Pereira thanks to the Fundação para a Ciência e Tecnologia, I. P (FCT), under the projects UIDB/04292/2020, UIDP/04292/2020, granted to MARE, and LA/P/0069/2020, granted to the Associate Laboratory ARNET.

Conflict of interest

The authors declare no conflict of interest.

Author details

Leonel Pereira[1*] and Miguel A. Pardal[2]

1 MARE—Marine and Environmental Sciences Centre/ARNET—Aquatic Research Network, Department of Life Sciences, University of Coimbra, Calçada Martim de Freitas, Coimbra, Portugal

2 CFE—Centre for Functional Ecology/TERRA Associate Laboratory, Department of Life Sciences, University of Coimbra, Calçada Martim de Freitas, Coimbra, Portugal

*Address all correspondence to: leonel.pereira@uc.pt

References

[1] Mimura N. Sea-level rise caused by climate change and its implications for society. Proceedings of the Japan Academy. Series B, Physical and Biological Sciences. 2013;**89**(7):281-301. DOI: 10.2183/pjab.89.281

[2] Wu C. "Central nation-peripheral barbarians in four directions-four seas": The geopolitical order of Land-Sea interactions of early Chinese civilization. In: Wu C, editor. The Prehistoric Maritime Frontier of Southeast China: Indigenous Bai Yue and Their Oceanic Dispersal. Singapore: Springer Singapore; 2021. pp. 3-24. DOI: 10.1007/978-981-16-4079-7_1

[3] Mogias A, Boubonari T, Realdon G, Previati M, Mokos M, Koulouri P, et al. Evaluating ocean literacy of elementary school students: Preliminary results of a cross-cultural study in the Mediterranean region. Frontiers in Marine Science. 2019;**6**:1-14. DOI: 10.3389/fmars.2019.00396

[4] Olson S, Jansen MF, Abbot DS, Halevy I, Goldblatt C. The effect of ocean salinity on climate and its implications for Earth's habitability. Geophysical Research Letters. 2022;**49**(10):e2021GL095748. DOI: 10.1029/2021GL09574

[5] Claudet J, Bopp L, Cheung WWL, Devillers R, Escobar-Briones E, Haugan P, et al. A roadmap for using the UN decade of ocean science for sustainable development in support of science, policy, and action. One Earth. 2020;**2**(1):34-42. DOI: 10.1016/j.oneear.2019.10.012

[6] The SA. "China seas" in world history: A general outline of the role of Chinese and east Asian maritime space from its origins to c. 1800. Journal of Marine and Island Cultures. 2012;**1**(2):63-86. DOI: 10.1016/j.imic.2012.11.002

[7] El-Abbadi M. Library of Alexandria. Encyclopedia Britannica. 2023. Available from: https://www.britannica.com/topic/Library-of-Alexandria [Accessed: February 04, 2023]

[8] Villiers AJ. James cook. In: Encyclopedia Britannica. 2023. Available from: https://www.britannica.com/biography/James-Cook [Accessed: February 04, 2023]

[9] Egerton FN. History of ecological sciences, part 37: Charles Darwin's voyage on the beagle. The Bulletin of the Ecological Society of America. 2010;**91**(4):398-431. DOI: 10.1890/0012-9623-91.4.398

[10] van Sebille E, Aliani S, Law KL, Maximenko N, Alsina JM, Bagaev A, et al. The physical oceanography of the transport of floating marine debris. Environmental Research Letters. 2020;**15**(2):023003. DOI: 10.1088/1748-9326/ab6d7d

[11] Gonzalez Cruz J, Sequera P, Molina Y, Picon R, Pillich J, Ghebreegziabhe AT, et al. 3.01 - climate and energy vulnerability in coastal regions: The case for US Pacific and northeast corridor coastal regions. In: Pielke RA, editor. Climate Vulnerability. Oxford: Academic Press; 2013. pp. 3-35. DOI: 10.1016/B978-0-12-384703-4.00302-6

[12] Jørgensen BB, Wenzhöfer F, Egger M, Glud RN. Sediment oxygen consumption: Role in the global marine carbon cycle. Earth-Science Reviews. 2022;**228**:103987. DOI: 10.1016/j.earscirev.2022.103987

[13] Olson S, Jansen MF, Abbot DS, Halevy I, Goldblatt C. The effect

of ocean salinity on climate and its implications for Earth's habitability. Geophysical Research Letters. 2022;**49**(10):e2021GL095748. DOI: 10.1029/2021GL095748

[14] Levin LA, Bett BJ, Gates AR, Heimbach P, Howe BM, Janssen F, et al. Global observing needs in the Deep Ocean. Frontiers in Marine Science. 2019;**6**:1-32. DOI: 10.3389/fmars.2019.00241

[15] Sarker S. Fundamentals of climatology for engineers: Lecture note. Eng. 2022;**3**(4):573-595. DOI: 10.3390/eng3040040

[16] Ocko IB, Ramaswamy V, Ming Y. Contrasting climate responses to the scattering and absorbing features of anthropogenic aerosol Forcings. Journal of Climate. 2014;**27**(14):5329-5345. DOI: 10.1175/JCLI-D-13-00401.1

[17] Irwin PGJ, editor. Dynamical processes. In: Giant Planets of our Solar System: An Introduction. Berlin, Heidelberg: Springer Berlin Heidelberg; 2003. pp. 133-196. DOI: 10.1007/3-540-37713-1_5

[18] Why does climate vary from one place to another. Available from: https://www.climate-policy-watcher.org/tropical-rainforest/why-does-climate-vary-from-one-place-to-another.html [Accessed: February 04, 2023]

[19] Dailidė R, Dailidė G, Razbadauskaitė-Venskė I, Povilanskas R, Dailidienė I. Sea-breeze front research based on remote sensing methods in coastal Baltic Sea climate: Case of Lithuania. Journal of Marine Science and Engineering. 2022;**10**(11):1779. DOI: 10.3390/jmse10111779

[20] Masuda K. Meridional heat transport by the atmosphere and the ocean: Analysis of FGGE data. Tellus A: Dynamic Meteorology and Oceanography. 1988;**40**(4):285-302. DOI: 10.3402/tellusa.v40i4.11801

[21] Davies AM, Xing J, Huthnance JM, Hall P, Thomsen L. Models of near-bed dynamics and sediment movement at the Iberian margin. Progress in Oceanography. 2002;**52**(2):373-397

[22] Kontakiotis G, Antonarakou A, Ruban DA. Geological oceanography: Towards a conceptual framework. Journal of Marine Science and Engineering. 2022;**10**(12):2027. DOI: 10.3390/jmse10122027

[23] Dickinson A, Gunn KL. The next decade of seismic oceanography: Possibilities, challenges and solutions. Frontiers in Marine Science. 2022;**9**. DOI: 10.3389/fmars.2022.736693

[24] Camargo JMR, Silva MVB, Ferreira Júnior AV, Araújo TCM. Marine geohazards: A bibliometric-based review. Geosciences. 2019;**9**(2):100. DOI: 10.3390/geosciences9020100

[25] Haigh ID, Pickering MD, Green JAM, Arbic BK, Arns A, Dangendorf S, et al. The tides they are a-Changin': A comprehensive review of past and future nonastronomical changes in tides, their driving mechanisms, and future implications. Reviews of Geophysics. 2020;**58**(1):e2018RG000636. DOI: 10.1029/2018RG000636

[26] Puerta P, Johnson C, Carreiro-Silva M, Henry L-A, Kenchington E, Morato T, et al. Influence of water masses on the biodiversity and biogeography of Deep-Sea benthic ecosystems in the North Atlantic. Frontiers in Marine Science. 2020;7. DOI: 10.3389/fmars.2020.00239

[27] National Research Council (US) Ocean Studies Board. 50 years of ocean

discovery: National science foundation 1950-2000. In: Achievements in Chemical Oceanography. Available from: https://www.ncbi.nlm.nih.gov/books/NBK208828/. National Academies Press (US): Washington (DC); 2000 [Accessed: February 04, 2023]

[28] Middelburg JJ, Soetaert K, Hagens M. Ocean alkalinity, buffering and biogeochemical processes. Reviews of Geophysics. 2020;**58**(3):e2019RG000681. DOI: 10.1029/2019RG000681

[29] Structure of Water. Available from: https://chem.libretexts.org/Bookshelves/Introductory_Chemistry/Introductory_Chemistry_(CK-12)/15%3A_Water/15.01%3A_Structure_of_Water [Accessed: February 04, 2023]

[30] Tornero V, Hanke G. Chemical contaminants entering the marine environment from sea-based sources: A review with a focus on European seas. Marine Pollution Bulletin. 2016;**112**(1):17-38. DOI: 10.1016/j.marpolbul.2016.06.091

[31] Lønborg C, Carreira C, Jickells T, Álvarez-Salgado XA. Impacts of global change on ocean dissolved organic carbon (DOC) cycling. Frontiers in Marine Science. 2020;7. DOI: 10.3389/fmars.2020.00466

[32] Weller RA, Baker DJ, Glackin MM, Roberts SJ, Schmitt RW, Twigg ES, et al. The challenge of sustaining ocean observations. Frontiers in Marine Science. 2019;**6**. DOI: 10.3389/fmars.2019.00105

[33] Bindoff N., Cheung WWL, Kairo JG, Arístegui J, et al. (eds.). Cambridge: Cambridge University Press; 2019. pp. 447-587. DOI: 10.1017/9781009157964.007.

[34] Gitz V, Meybeck A, Lipper L, Young C, Braatz S. Climate Change and Food Security: Risks and Responses. Rome, Italy: FAO; 2016

[35] Sigman DM, Hain MP. The biological productivity of the ocean. Nature Education Knowledge. 2012;**3**(10):21. Available from: https://earth-system-biogeochemistry.net/wp-content/uploads/2021/05/Sigman_and_Hain_2012_NatureEdu.pdf [Accessed: February 04, 2023]

[36] Alberts B, Johnson A, Lewis J, et al. Molecular Biology of the Cell. 4th ed. New York: Garland Science; 2002. Chloroplasts and Photosynthesis. Available from: https://www.ncbi.nlm.nih.gov/books/NBK26819/ [Accessed: February 04, 2023]

[37] Pereira L, Correia F. Algas Marinhas da Costa Portuguesa - Ecologia, Biodiversidade e Utilizações. Paris: Nota de Rodapé Editores; 2015. 341 pp. ISBN 978-989-20-5754-5

[38] Dokulil MT, Qian K. Photosynthesis, carbon acquisition and primary productivity of phytoplankton: A review dedicated to Colin Reynolds. Hydrobiologia. 2021;**848**(1):77-94. DOI: 10.1007/s10750-020-04321-y

[39] Hamner WM. Predation, cover, and convergent evolution in epipelagic oceans. Marine and Freshwater Behaviour and Physiology. 1995;**26**(2-4):71-89. DOI: 10.1080/10236249509378930

[40] Pereira L, Gonçalves AMM, editors. Plankton communities. In: Open Access Peer-Reviewed Edited Book. London, UK, London, UK: Intech Open; 2022. DOI: 10.5772/intechopen.91092

[41] Wetzel RG. 16 - planktonic communities: Zooplankton and their interactions with fish. In: Wetzel RG, editor. Limnology. 3rd ed. San Diego: Academic Press; 2001. pp. 395-488

[42] Heinrichs ME, Mori C, Dlugosch L. Complex interactions between aquatic organisms and their chemical environment elucidated from different perspectives. In: Jungblut S, Liebich V, Bode-Dalby M, editors. YOUMARES 9 - The Oceans: Our Research, our Future: Proceedings of the 2018 Conference for YOUng MArine RESearcher in Oldenburg, Germany. Cham: Springer International Publishing; 2020. pp. 279-297

[43] Araújo GS, Pacheco D, Cotas J, JWAD S, Saboya J, Moreira RT, et al. Plankton: Environmental and economic importance for a sustainable future. In: Pereira L, Gonçalves AM, editors. Plankton Communities. London: IntechOpen; 2022. DOI: 10.5772/intechopen.100433

[44] Costello M, Chaudhary C. Marine biodiversity, biogeography, Deep-Sea gradients, and conservation. Current Biology. 2017;**27**:R511-RR27. DOI: 10.1016/j.cub.2017.04.060

[45] Pinel-Alloul B, Ghadouani A. Spatial heterogeneity of planktonic microorganisms in aquatic systems. In: Franklin RB, Mills AL, editors. The Spatial Distribution of Microbes in the Environment. Dordrecht: Springer Netherlands; 2007. pp. 203-310. DOI: 10.1007/978-1-4020-6216-2_8

[46] Mora C, Tittensor DP, Adl S, Simpson AG, Worm B. How many species are there on earth and in the ocean? PLoS Biology. 2011;**9**(8):e1001127. DOI: 10.1371/journal.pbio.1001127

[47] Wasmund N, Siegel H, Bohata K, Flohr A, Hansen A, Mohrholz V. Phytoplankton stimulation in frontal regions of Benguela upwelling filaments by internal factors. Frontiers in Marine Science. 2016;**3**. DOI: 10.3389/fmars.2016.00210

[48] Lindemann C, Backhaus JO, St. John MA. Physiological constrains on Sverdrup's critical-depth-hypothesis: The influences of dark respiration and sinking. ICES Journal of Marine Science. 2015;**72**(6):1942-1951. DOI: 10.1093/icesjms/fsv046

[49] Mackey KRM, Paytan A. Phosphorus cycle. In: Schaechter M, editor. Encyclopedia of Microbiology. 3rd ed. Oxford: Academic Press; 2009. pp. 322-334. DOI: 10.1016/B978-012373944-5.00056-0

[50] Cavan EL, Belcher A, Atkinson A, Hill SL, Kawaguchi S, McCormack S, et al. The importance of Antarctic krill in biogeochemical cycles. Nature Communications. 2019;**10**(1):4742. DOI: 10.1038/s41467-019-12668-7

[51] Aldemaro R. When whales became mammals: The scientific journey of cetaceans from fish to mammals in the history of science. In: Aldemaro R, Edward OK, editors. New Approaches to the Study of Marine Mammals. Rijeka: IntechOpen; 2012 1

[52] Davis RW. A review of the multi-level adaptations for maximizing aerobic dive duration in marine mammals: From biochemistry to behavior. Journal of Comparative Physiology. B. 2014;**184**(1):23-53. DOI: 10.1007/s00360-013-0782-z

[53] Favilla AB, Costa DP. Thermoregulatory strategies of diving air-breathing marine vertebrates: A review. Frontiers in Ecology and Evolution. 2020;**8**. DOI: 10.3389/fevo.2020.555509

[54] Jung S, Chau TV, Kim M, Na W-B. Artificial seaweed reefs that support the establishment of submerged aquatic vegetation beds and Facilitate Ocean macroalgal afforestation: A review. Journal of Marine Science and Engineering. 2022;**10**(9):1184. DOI: 10.3390/jmse10091184

[55] Rotter A, Barbier M, Bertoni F, Bones AM, Cancela ML, Carlsson J, et al. The essentials of marine biotechnology. Frontiers in Marine Science. 2021;**8**. DOI: 10.3389/fmars.2021.629629

[56] Blunt JW, Copp BR, Munro MHG, Northcote PT, Prinsep MR. Marine natural products. Natural Product Reports. 2004;**21**(1):1-49. DOI: 10.1039/b305250h

[57] Lindequist U. Marine-derived pharmaceuticals - challenges and opportunities. Biomolecules & Therapeutics. 2016;**24**(6):561-571. DOI: 10.4062/biomolther.2016.181

[58] Song JH, Murphy RJ, Narayan R, Davies GB. Biodegradable and compostable alternatives to conventional plastics. Philosophical Transactions of the Royal Society of London. Series B, Biological Sciences. 2009;**364**(1526):2127-2139. DOI: 10.1098/rstb.2008.0289

[59] Santos VP, Marques NSS, Maia PCSV, Lima MAB, Franco LO, Campos-Takaki GM. Seafood waste as attractive source of chitin and chitosan production and their applications. International Journal of Molecular Sciences. 2020;**21**(12):4290. DOI: 10.3390/ijms21124290

[60] López-Hortas L, Flórez-Fernández N, Torres MD, Ferreira-Anta T, Casas MP, Balboa EM, et al. Applying seaweed compounds in cosmetics, cosmeceuticals and Nutricosmetics. Marine Drugs. 2021;**19**(10):552. DOI: 10.3390/md19100552

[61] Pereira L. Colloid producing seaweeds. In: Kim S-K, editor. Encyclopedia of Marine Biotechnology. New Jersey, USA: Wiley; 2020. pp. 161-326. DOI: 10.1002/9781119143802.ch8

[62] Lomartire S, Gonçalves AMM. Novel Technologies for Seaweed Polysaccharides Extraction and Their use in food with therapeutically applications - A review. Food. 2022;**11**(17):2654. DOI: 10.3390/foods11172654

[63] Bošnjaković M, Sinaga N. The perspective of large-scale production of algae biodiesel. Applied Sciences. 2020;**10**(22):8181. DOI: 10.3390/app10228181

[64] Sarwer A, Hamed SM, Osman AI et al. Algal biomass valorization for biofuel production and carbon sequestration: A review. Environmental Chemistry Letters. 2022;**20**:2797-2851. DOI: 10.1007/s10311-022-01458-1

[65] Takeda H, Yoneyama F, Kawai S, Hashimoto W, Murata K. Bioethanol production from marine biomass alginate by metabolically engineered bacteria. Energy & Environmental Science. 2011;**4**(7):2575-2581. DOI: 10.1039/C1EE01236C

[66] Ayilara MS, Babalola OO. Bioremediation of environmental wastes: The role of microorganisms. Frontiers in Agronomy. 2023;**5**. DOI: 10.3389/fagro.2023.1183691

[67] Crawford RL. Bioremediation. In: Dworkin M, Falkow S, Rosenberg E, Schleifer K-H, Stackebrandt E, editors. The Prokaryotes: Volume 1: Symbiotic associations, Biotechnology, Applied Microbiology. New York, NY: Springer New York; 2006. pp. 850-863. DOI: 10.1007/0-387-30741-9_26

[68] Bisht S, Pandey P, Bhargava B, Sharma S, Kumar V, Sharma KD. Bioremediation of polyaromatic hydrocarbons (PAHs) using rhizosphere technology. Brazilian Journal of Microbiology. 2015;**46**. DOI: 10.1590/S1517-838246120131354

Section 2

Aquaculture and Biodiversity

Chapter 2

Seaweed Aquaculture Importance in Sustainable Economy in an Era of Climate Change

Madalena Mendes, Alina Hillinger, Kay Ihle, Miguel Cascais, Pedro Andrade, João Cotas, Diana Pacheco, João Carlos Marques, Leonel Pereira and Ana Marta Mendes Gonçalves

Abstract

Seaweeds have been used globally for different purposes. Recent developments in technology coupled with an increasing interest in the resource have created a need for seaweed aquaculture to meet the demand. This review tackles the question of whether seaweed aquaculture has the potential to support the blue economy as well as climate change adaptation. Seaweed aquaculture represents an economically viable sector that has the potential to support the blue economy. The sector succeeds in meeting sustainability requirements while supporting human health and wealth through qualities such as naturally acting as a bio-filter, reducing ocean acidification, carbon sequestration, provision of habitat, and coastal protection; seaweed aquaculture can be used as a tool in conservation and climate adaptation. However, seaweed aquaculture is still in its infancy in many areas of the world, especially in the West, and there is a need for improved legislation and guidelines. Furthermore, several environmental hazards, such as physical stress (abiotic and abiotic factors), genetic depletion, and the introduction of non-native species, can cause future problems, particularly if legislation restrictions are not fully revised to ensure the safety of seaweed aquaculture. The sector poses great opportunities and is a sustainable way forward that is necessary to meet increasing demand.

Keywords: aquaculture, blue carbon, blue economy, climate change adaptation, macroalgae, sustainability

1. Introduction

Seaweed has been used by humans for centuries for different purposes such as nutrition, medicine, and fertilizer [1–3]. This group of marine plants, also referred to as macroalgae, is composed of three taxonomic groups: brown (Ochrophyta, Phaeophyceae), red (Rhodophyta), and green (Chlorophyta) seaweeds [4, 5]. They are distributed horizontally in different zones, which are supratidal (supralittoral), intertidal (littoral), and subtidal (sublittoral) regions of the seas and oceans [4].

Green, brown, and red seaweeds are generally distributed in the intertidal, tidal, or upper subtidal zone and subtidal regions, respectively [6].

The presence of seaweeds in all climate belts, great biodiversity, their compositional abundance of vitamins and macro- and micronutrients, and the fact that most seaweeds are edible suggest that seaweeds have played an important role as human food during human evolution. Seaweeds have served as a cheap and easily accessible crop in the daily fare for coastal populations, especially in Southeast Asia, where seaweeds have for millennia been considered as valuable sea vegetables. In the face of an ever-growing population, seaweeds are predicted to play an increasing role as a food source. Based on their fast growth, [6] global availability and suitability as food, they represent a viable possibility in providing food security [7].

Nowadays, 31–38% of the seaweed aquaculture biomass is directly used as food [8]. Polysaccharides extraction is the second largest application of seaweeds [8, 9] being commercially employed in food products due to their jellifying qualities, as food additives, other food supplements, and hydrocolloid compounds for medicines, nutraceuticals, and cosmetic products [8, 9]. Agar, alginate, and carrageenan are economically exploited [10, 11]. Seaweeds are also used in applications including animal feed [12], fertilizers, bioplastics, and biofuel [8, 9].

Currently, most of all commercial seaweed comes from aquaculture rather than wild harvest [13]. Although the seaweed market is still emerging in many parts of the planet, the established market is already considerable [14]. To support growing seaweed market resource demands while ensuring high-quality seaweed products, an expansion of seaweed aquaculture is required [11]. This expansion holds a multitude of benefits. It offers the opportunity to increase an economically potent market while protecting wild populations from overexploitation through harvesting [11]; it supports a sustainable market [15] while algae-based products have the potential to replace unsustainable products [8, 9] and food production methods with sustainable alternatives [16].

This chapter elaborates upon the current state of seaweed aquaculture globally and cultivation methods that are commonly applied. Furthermore, it discusses the potential benefits and drawbacks of seaweed aquaculture regarding an emerging blue economy and its wide-ranging effects on society and the environment. Lastly, the ecosystem services and climate change adaptation potential of seaweed cultivation are discussed.

2. Seaweed aquaculture: current status

2.1 Global seaweed aquaculture

Seaweeds have vast applications for human benefit, with their full potential still being explored. In 2020, the total production of world seaweed aquaculture grew by half a million tons, up by 1.4% from 34.6 million tons of fresh weight (FW) in 2019 [17]. Over 99% of the production is grown by various methods in Asia. Typically, only five major countries (China, Indonesia, the Philippines, Korea, and Japan) produce over 97% of seaweed globally [18]. Amounting to more than 32 million tons (Mt) in 2017, seaweed production is responsible for a large proportion of the global aquaculture's liveweight, only topped by freshwater fish [8, 13]. Within the global aquaculture production, the two most cultivated species groups in terms of fresh weight are seaweeds: Japanese kelp (family Laminariaceae) and *Eucheuma* spp. [8].

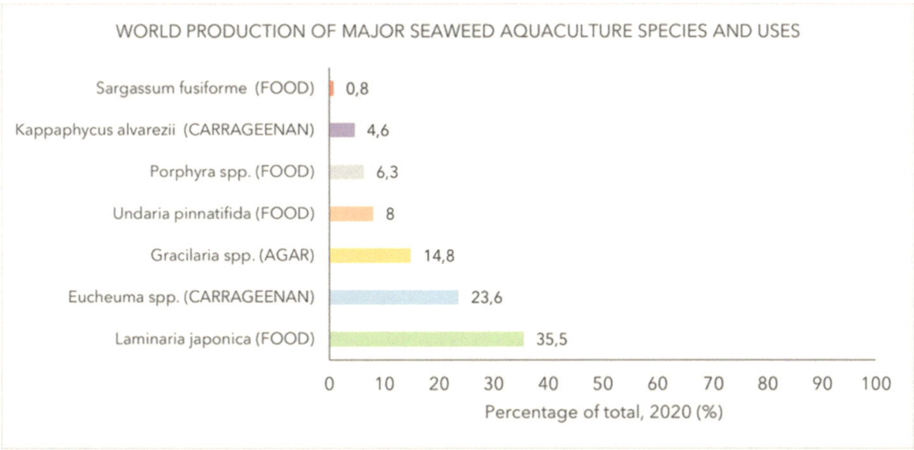

Figure 1.
Major seaweed aquaculture species and their uses.

Since seaweeds have been a part of the cuisine of many Asian countries for about for hundreds of years, it is not surprising that seaweed production is dominated by Asian countries [19] and 99% of seaweed aquaculture is situated in Asia [8]. In contrast, in Europe and the Americas, seaweeds are still mostly harvested from nature [7, 20, 21]. Outside of Asia, only Zanzibar's and Chile's aquacultures produce significant amounts of seaweed on a global scale [7]. In Europe, only 1% of the produced seaweed biomass consists of aquaculture, of which Norway, Denmark, and Spain are at the forefront of the number of companies [20]. **Figure 1** presents the major seaweed aquaculture species and their uses [17].

2.2 Seaweed aquaculture cultivation methods

Cultivation methods can be characterized by their location, approach, farming process, or cultivation technique [7, 22, 23]. Some of these (**Figure 2**) will be discussed here with special regard to their sustainability.

Offshore and nearshore cultivation are terms that differentiate between farms far and close to the coast, although an official legal definition does not exist [22]. Nearshore aquaculture is the most common seaweed farming method and has the benefits of needing little labor and seaweeds being quite sheltered against harsh conditions of the open water [22]. The environmental upside of this cultivation technique is that it can take up nutrients from agricultural effluent [24], acting as a nutrient bioextraction system [19]. For large seaweed farms, interest in offshore cultivation has been growing [22]. It is slightly more labor-intensive than nearshore cultivation but still inexpensive [22]. However, seaweeds are exposed to harsh conditions offshore and are faced with several other difficulties related to epiphytes, making the method only suitable for certain seaweed species [22]. The problem with epiphytes is that they compete with macroalgae for space, nutrients, light, etc., causing detachment of host tissue. It can be environmentally friendly, but the downside is that the transport to and from the cultivation system adds to the energy usage [22].

In contrast to the previous two methods, onshore cultivation takes place in a closed or semi-closed system. This enables much control over abiotic factors, like seawater in- and outflow, nutrients, light, pH, CO_2, and salinity, and protects them from harsh

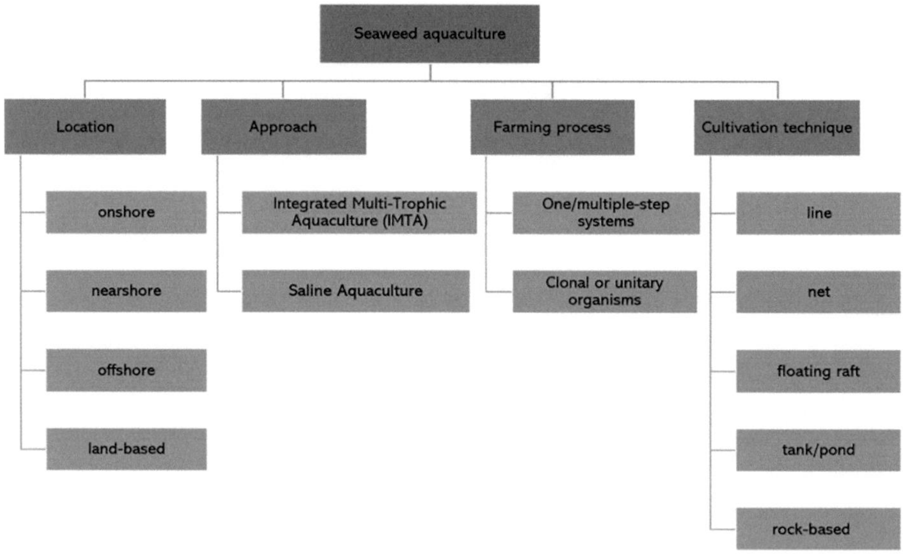

Figure 2.
Cultivation methods assessment based on several parameters (location, approach, farming process, or cultivation technique).

environmental conditions like winds and waves [22, 25]. Although expensive, it enables consistent production, which makes it attractive for commercial large-scale production, especially for high-value product extraction [22, 25]. Besides, it is suitable for a wide range of different species that would be difficult to cultivate otherwise [25]. A specific form of onshore land-based cultivation is Saline Aquaculture, which makes use of saline groundwater [22]. Earthen or plastic-lined ponds, raceways, and tanks are some of the farming systems used in this method. Some advantages of marine algae culture include the ability to use existing agricultural farms where saline water is available, which is less constrained by the need for additional resources than farming in the sea. Another advantage of cultivating marine algae in island saline water (ISW) is that it can provide an additional source of income and raw marine algae for the aquaculture seaweed industry and has a lower investment than offshore and nearshore farming [22], which could make it an interesting option for lower-income countries. Concerning sustainability, there are both advantages and disadvantages to onshore aquaculture of seaweeds. Since the effluent is controlled, the impact of onshore cultivation on its physical environment can be minimized, even when nutrients are added [25]. Additionally, because onshore systems are three-dimensional, they are more efficient in terms of area than classic offshore cultivation [25]. However, onshore cultivation comes with significant energy requirements [25]. Moreover, in contrast to offshore cultivation, it is located on land and thus lacks the advantage of offshore cultivation of not competing for arable land [22].

Especially of interest for conserving the environment is integrated multi-trophic aquaculture (IMTA). This cultivation method uses species of different trophic levels [26]. More specifically, it combines waste or nutrient-producing aquaculture with extractive species like seaweeds, thus mitigating the impact of the system on its environment [19, 26]. For example, in Portugal, sustainable production of seaweed (Algaplus company) uses this IMTA concept, farming seaweeds in proximity to several species at different trophic levels, allowing the reduction of aquaculture wastes [27]. IMTA is practiced in a controlled environment on land, with organic certification for quality, traceability, stability of supply, and a small carbon footprint. Although it still faces some challenges regarding contaminants and complexity, it is a promising "ecological approach to aquaculture" and is part of the EU's Blue Growth Strategy [26]. In general, seaweeds are cultivated in one or multiple-step farming systems [28]. Clonal species, such as *Gracilaria* and *Kappaphycus* (Rhodophyta), can be fragmented vegetatively and propagated directly for growth in culture systems. This can be done at different scales, moving from intensive on-land tanks or ponds to extensive open sea culture systems using long lines or rafts. On the other hand, propagation of unitary seaweeds such as kelp for industrial cultivation requires a hatchery/nursery. The different cultivation systems have been thoroughly tested experimentally, but the most commercially successful systems have been those that culture seaweeds at sea due to lower operational and capital costs [7].

3. Seaweed aquaculture: a future road to blue economy?

In recent years, the concept of the 'blue economy' has attracted the attention of policymakers, managers, and scientists alike. Since the introduction of the term to protect the global seas and their resources while supporting the human population [20], current definitions of blue economy have progressed to encompass a vast array of components far beyond its original scope. For example, according to the United Nations, blue economy is "a long-term strategy aimed at supporting sustainable economic growth through oceans-related sectors and activities, while improving human well-being and social equity and preserving the environment" [29]. According to the World Bank, the blue economy is the "sustainable use of ocean resources for economic growth, improved livelihoods, and jobs while preserving the health of ocean ecosystem" [30]. European Commission defines it as "All economic activities related to oceans, seas and coasts. It covers a wide range of interlinked established and emerging sectors" [31]. Nowadays, blue economy describes an approach to the immediate and long-term development and growth of a sustainable economy [32, 33], which promotes human health and wealth, food security, and the conservation of the marine environment and its resources [33–35]. Forms of blue economy have been applied in traditional management for millennia [36], as exemplified by Keen et al. for Pacific islanders [33]. This concept is also closely related to the term "phyconomy," which is hereby coined to describe a general concept that embraces large-scale, sustainable seaweed farming for economic benefit in coastal waters. As an emerging sector that holds the potential for vast economic growth in upcoming decades [22], the aquaculture cultivation of seaweeds is often associated with the concept of blue economy. This association is reflected in notions such as the "blue dimensions of aquaculture" [37], "blue bioeconomy" [13, 20], or "phyco-economy" (a concept of the industrialized and sustainable cultivation of seaweeds) [38]; which already tie seaweed aquaculture to the blue economy. In the following paragraphs, the potential of seaweed aquaculture to the blue economy will be elaborated upon, in respect to aims of blue economy: human health and wealth, food security, promotion of social factors as well as the conservation of the marine environment and resources.

3.1 Economy

The industry has a large potential for fast expansion, as initial investments are comparatively low (depending on species, technique, and location), while returns are high [33]. The time until settlement, however, is very short: the period between planting and the first harvest of seaweeds spans less than 2 months [11, 33].

Additionally, areas suited for seaweed aquacultures amount to 48 million km^2, over 99% of which are currently unused for the purpose [34, 39, 40]. Less than a third of the 132 countries with suitable areas currently engage in seaweed cultivation [34, 39]. Seaweed farming thus has a high potential for expansion of the sector, which is supported by a growing relevance of seaweed products [22], while low financial thresholds offer an opportunity for economic development and the creation of jobs in low-income countries. It is worth mentioning that previous studies that have identified "high opportunity marine ecoregions" (HOMEs) for the development of seaweed aquaculture in consideration of socio-economic and environmental factors have determined that most HOMEs border wealthy countries rather than low-income countries [15]. Again, this, if not considered, could further add to global inequalities.

Seaweed-based products have several valuable economic applications, including seaweeds for consumption, dried seaweed, and seaweed extracts, as well as products based on these applications. Fresh or dried seaweeds for human consumption, hereafter referred to as edible seaweeds, have a higher market value than products for extraction purposes [10]. Edible seaweeds currently render >3 billion US$ annually [41]. Extracted algal polysaccharides (namely fucoidan, alginate, ascophyllan, laminarin, porphyrans, agar, carrageenans, and ulvans) have high market value already and are expected to become increasingly demanded and thus economically prized [33]. Agar and carrageenan (both retrieved from Rhodophyta, red seaweeds) [42] as well as alginate (retrieved from Phaeophyceae, brown seaweeds) are presently the most demanded algal polysaccharides [33, 43]. By 2024 [42, 44], agar, carrageenan, and alginate are expected to reach a market value of 1 billion US$, with growth continuing until 2025 [33]. On the global market, a diversion between market value and demand is regionally present: in several Asian countries, edible seaweeds are highly demanded for consumption, while in wealthy westernized countries, the market offers a large niche for seaweed extract products [10]. It is also relevant to note that while the interest and approximated demand in the use of seaweeds for biofuel is growing, its economic value is relatively low [7].

Recent data on the growth of the sector is ambiguous. Seaweed aquaculture currently provides annual revenues of around 6 billion US$ [41, 45], and with a threefold increase in value in the past 100 years as well as continued growth until the last decade [10], seaweed cultivation is regarded as a promising economic sector by a multitude of publications [22]. A recent study has found that in the past 10 years, a 150% increase in "algae-producing companies" has taken place [20]. However, recent reports have met this enthusiasm with doubt. Buschmann et al. [7] have regarded the potential of the seaweed aquaculture sector as "larger than its actual scale," historically as well as at current. In a 2011 report, the Food and Agriculture Organization (FAO) estimated the annual value of seaweed at around 7 billion US$, which indicates a 1 billion US$ decrease between 2011 and 2018 [41, 46]. Moreover, in 2020, the FAO illustrated that the production of cultivated algae, a majority of which are seaweeds, has encountered low growth in recent years, with a decrease of 0.7% in 2018 [13]. While production trends do not necessarily reflect the market, they support Buschmann et al. [7] in their reservation about the potential of the sector at its development. Additionally, repeated mentioning of "hurdles," such as unclear legislation, protocols, and guidelines, and contradiction within the aim of the blue economy itself, have been brought forth in recent publications on the subject [7]. Still, most studies conclude that seaweed aquaculture has a high potential within the scope of the blue economy once current hurdles are overcome [7, 20].

3.2 Human health and wealth

An expansion of the seaweed aquaculture industry also offers employment and a linked increase in socio-economic wealth in low-income countries, as has already been recorded in several such countries [7, 14, 22, 47]. The industry also offers safe work conditions as the aquaculture is located in the shallow intertidal [11]. Ferdouse et al. [34] have pointed out that seaweed aquaculture has the potential to especially support and empower women in areas where they have thus far had little independence, which can support their families with their own income [34]. However, this argument of moving toward gender equality should not be treated lightly, and its use as a "catchphrase" should be avoided, as the curation of jobs does not have a direct positive effect on minorities or currently oppressed groups. The same paper also included children in this consideration of safe access to the aquaculture industries [11]. Again, caution is urged, and a need for the provision of support and guidance presents itself. An increased cultivation of edible algae, which currently represents around 40% of the worldwide seaweed production, has the potential to reduce food shortage [7, 33]. Again, this especially applies in low-income countries, as it is a cheap and efficient alternative to land-based agriculture food production. Seaweeds, often also referred to as sea vegetables, have the potential to significantly add to the global protein supply within the next decade [7]. Lastly, an interconnection inside the formation interconnection between non-market value ecosystem services and human welfare, which adds to the objective of the blue economy, has been established [7].

3.3 Conservation of the marine environment and resources

Seaweed's aquaculture cultivation is frequently described as sustainable and environmentally friendly and thus fits well into the framework of the blue economy. This characteristic is based on several factors. Firstly, unlike terrestrial agriculture, the farming of seaweeds does not utilize any fertilizer and pesticides [34] or feed and other organic waste compared to fish monoculture [6, 48]. As such, it does not produce nutrient-rich, and consequently, eutrophication causes discharges. On the contrary, seaweeds absorb excessive nutrients such as nitrogen, carbon, and phosphorus from the water column and reduce associated eutrophication [6, 19, 49]. Furthermore, wastewater discharges can be filled with heavy metals and other pollutants, and to prevent these compounds from entering the food chain, seaweeds, with their biosorption properties, can be taken advantage of [50]. Due to the macromolecules (mainly polysaccharides) in the cell walls, seaweeds can bind metal ions and heavy metals such as arsenic or lead [50, 51]. However, seaweeds that have bio-accumulated a high concentration of heavy metals need to be removed so they do not endanger the food chain [52] and cannot be used for human consumption [51]. Seaweed aquacultures can accordingly be a tool in bioremediation and reduce eutrophication [15, 53]. It is relevant to note that the bioremediation efficiency differs among species [54]. The energy transfer from dissolved nutrients to seaweed biomass also contributes to an increase in food availability for associated species [55]. Beyond that, as mentioned previously, seaweed cultivation can be linked to fish raising in integrated multi-trophic aquaculture (IMTAs) [6, 48]. In IMTAs, the excess nutrient discharge (organic and inorganic substances) from fish production systems can be absorbed and utilized by seaweeds, which act as biofilters and turn into biomass production [6, 19, 56–58]. Through creating a link between the rearing of live animals and seaweeds, the blue economy is supported beyond the direct extent of seaweed aquaculture. Instead, the seafood and fish aquaculture sectors, which are heavily criticized for their environmental impact while holding the key to providing sufficient seafood and conserving wild fish populations, can be supported [59, 60]. Additionally, by replacing wild fish with seaweed components in fish feed [61, 62], wild fish populations can be protected.

Aquaculture cultivation of seaweed also forms a temporary habitat for other potentially vulnerable species [20, 63, 64]. Increases in biodiversity of herbivores as well as other species attracted to the complex habitat such as fish and invertebrates are commonly observed [6, 63]. Seaweed farms can also act as nursing habitats [65], as exemplified by Hughes et al. [66]. However, these positive effects on biodiversity are limited to the period until harvest [63]. Apart from habitat provision, the physical structure of seaweed habitats presents several beneficial services to the surrounding ecosystems [34]. It depletes waves and currents of their energy before meeting the coast and concurrently prevents erosion [63, 67].

According to Choudhary et al., commercial seaweed cultivation can aid in the conservation and recovery from overexploitation of affected macroalgae species, as they are released from unsustainable harvesting pressure. An example is *Gracilaria* sp. (Rhodophyta), which was impacted by unsustainable harvesting in Chile and has been released of this pressure [59, 60].

Similarly, commercial species such as *Laminaria hyperborea* (Ochrophyta, Phaeophyceae) in Norway are currently harvested by specialized trawling gear [6]. The harvest of seaweeds by trawling implies the removal of benthic flora and fauna as well as the destruction of habitats [6]. Expanding the cultivation of such species offers the potential to meet the increasing demand for the product [6] and release affected ecosystems from the pressure of trawling.

Seaweed aquaculture offers a variety of opportunities to support the adjacent ecosystems, the effects of which are often described as ecosystem services. As such, it generates an extended non-market value [7]. Ecosystem services are variable and include supporting, regulating, and provisioning services [49]. While the value of these non-market-value services is difficult to estimate, attaching a monetary value to these services has the potential to further enhance the economic value of seaweed aquaculture beyond their marketable value [68].

A comprehensive analysis considering where each service of seaweed cultivation can be placed within the paradigm of ecosystem services is provided by Hasselström et al. [49].

However, several questions arise regarding the potential negative effects of seaweed aquaculture on the ecosystem. Does future commercialized extensive cultures of seaweeds pose a risk to the genetic variation of wild populations? Could they mirror land-based agricultural monocultures in their negative effects on the environment?

Firstly, the environmental and physical impacts of seaweed aquaculture systems need to be considered. Depending on the species and system, the mariculture of seaweeds can require the clearing of an area from rocks or species growing in the area. When floating structures are used, they can cause shading of the adjacent water column (**Figure 3**). All systems can cause a change in sedimentation, flow regime,

Figure 3.
Chondrus crispus aquaculture floating system.

and other disturbances through the swaying of the seaweeds or even indirectly through an altered species community surrounding the systems [63]. In land-based systems, problems are fewer. A main issue is the use of space for pond creation [22]. Additionally, seawater must be supplied and then discharged. However, the discharge has few excess nutrients, and associated problems are expected to be minor [63]. Spatial planning can be crucial here, and assigning areas for aquacultures can cause or be hindered by conflicts of interest.

A vast proportion of cultured species is not native to the region of cultivation, such as the red seaweed *Gracilaria vermiculophylla* (formerly *Agarophyton vermiculophyllum*), native from Japan, that has invaded estuaries throughout the whole world, and the brown seaweed *Sargassum muticum* (Phaeophyceae), native seaweed from Japan and China, currently widespread along the European shoreline [69]. These non-native species have historically caused invasions in the range of introduction, where they can compete with or even outcompete native species [70]. As invasive species become dominant, they can alter the local species assemblage [70] and influence the functioning and stability of the ecosystem [6]. Furthermore, the introduction of seaweeds to an area for aquaculture increases the likelihood of unintentional introduction of non-indigenous "hitchhiker" species, including pathogens [22]. All of these non-native species can spread easily over short and also longer distances due to the open and dispersive nature of the ocean.

Next to non-native species, the so-called crop-to-wild gene flow negatively affects local seaweed populations. Selective breeding of the cultivars used for aquaculture crops is practiced to enhance beneficial traits including biomass, growth rate [6], or disease resistance [19]. This practice tends to reduce the genetic variability and associated evolutionary adaptability of the cultivated seaweed [19]. Again, the dispersive and open nature of marine systems allows for gene flow between wild and cultivated crops, through which the selected alleles can become established in wild populations [70]. The resulting genetic depletion can leave the native population vulnerable, as its ability to adapt to a changing environment is decreased [6, 63, 71].

4. Seaweed aquaculture and climate change adaptation

While seaweed aquaculture carries great potential as a sector in the blue economy based on its primarily positive impacts on the above-described factors, it exceeds the current definition of blue economy by the United Nations. Large-scaled seaweed cultivation can also positively affect climate change scenarios and allow for climate change adaptation. This fact scenario will be discussed here.

Currently, increased emissions of the acidic gas carbon dioxide induce an increase of CO_2 in the atmosphere and oceans. This increase leads to the acidification of global oceans. Ocean acidification is especially harmful to calcified seaweeds (such as Corallinaceae) and animals with calcified body structures, as the decreased pH dissolves these structures [72]. Seaweeds are autotrophic plants and, as such, perform photosynthesis. Through this process, they transform CO_2 into energy [73, 74]. Seaweeds continuously take up CO_2 and in turn reduce its concentration in the oceans [45, 75], which consequently increases the pH of seawater and reduces ocean acidification [39, 76]. As a consequence, seaweed beds, including aquacultures, provide refuge to vulnerable calcifiers and other species [72, 75, 77]. Seaweed production through cultivation is crucial for the capture of CO_2 from the atmosphere [73]. The use of seaweed as a CO_2 sink provides one way to extend blue carbon strategies [78, 79]. Recent studies found that macroalgae detritus has a high potential for carbon sequestration [80, 81]. Unlike other ecosystems that provide plants, like seagrass, seaweed does not form sediments around their holdfasts due to the hard nature of the substrate they grow on [80, 82]. However, seaweed detritus is transported to recipient sites, where they are deposited along with the absorbed atmospheric carbon [80]. Such recipient sites were recently found to extend coastal areas by up to 5000 km and include the deep seas [83]. In offshore and deep areas, limited human influence and wave action reduce the probability of the carbon returning to the atmosphere and instead remain sunk [82]. Here, carbon provides an important food source, and on the other hand, it is sunk in the sediments of the deep coastal benthic food web [81, 84]. It has been estimated that over half of the green carbon capture, thus the capture of CO_2 through terrestrial and marine systems, is through the marine realm [81]. Mashoreng et al. [85] have found that an average of 57.64 tons $CO_2\ ha^{-1}\ yr^{-1}$ in ocean-based seaweed aquacultures and 12.38 tons $CO_2\ ha^{-1}\ yr^{-1}$ in land-based aquacultures could be sequestered. In strategies toward blue carbon capture, it should thus be considered that land-based pond systems offer less potential for blue carbon capture [85]. Additionally, it was found that seaweeds at different developmental stages display different carbon sequestration potentials [86]. Fakhraini et al. [86] determined that in research aimed at analyzing carbon sequestration potential on seaweed *Kappaphycus striatus* adult plants have 32.78% more potential than young plants. While the potential of seaweeds in climate change and their application in "Blue carbon farming" is vast, it is important to consider the spatial planning of seaweed farms. Here, it is especially important to assess and determine the minimum and maximum extent of the farms to ensure efficiency and prevent depletion of coastal carbon. A depletion of carbon could affect other ecosystems, such as seagrass meadows, which in turn provide habitat and resources to other organisms [75]. While the use of seaweed aquacultures in the blue carbon framework is still relatively undeveloped, and knowledge gaps are present, it has been determined to play an imperative role in carbon capture [87].

While transforming CO_2 into energy, seaweeds produce O_2. Another effect of climate change is eutrophication, a removal or lack of oxygen predominantly from coastal areas [88]. Eutrophication is the enrichment of water with nutrients such as minerals, nitrogen, and phosphorus-containing materials. This frequently leads to the unwanted and excessive growth of aquatic or marine plants (algae blooms). Eutrophication can occur naturally, but it can also be accelerated by allowing water rich in dissolved fertilizers to seep into nearby lakes and streams or by the introduction of sewage effluent into rivers and coastal waters. Seaweeds can be used to reduce the nitrogen and phosphorus content of effluents from sewage treatment. Many seaweeds prefer taking up ammonium as the form of nitrogen for their growth, and ammonium is the prevalent form of nitrogen in most domestic and agricultural wastewater. Seaweeds add O_2 to the ecosystem and can help mitigate this issue. Seaweed aquaculture has the potential to act as a supplementary force to wild seaweed populations in their ability to perform such services [75]. Additionally, seaweeds grown in coastal aquacultures have the potential to support other climate change adaptations—strategies to mitigate the effects of climate change (events) on the planet and human societies [75, 89].

Seaweed aquaculture, however, also has less direct and apparent potential in climate change mitigation. Seaweed can be used directly as a replacement for traditional fossil fuels. Several species of seaweeds accumulate high levels of carbohydrates, which are suitable as substrates for microbial conversion processes, e.g., for the production of bioethanol, biobutanol as biofuels, or other desirable chemicals with an attractive high product value. Kelp species contain 50–60% carbohydrates of the dry weight, and cultivation techniques have been firmly established for the last 50 years. Moreover, kelp is cultivated in large quantities, providing an abundant and potential carbon-neutral renewable resource with the potential to reduce greenhouse gas (GHG) emissions and the man-made impact on climate change [90].

Another application of macroalgae is in agriculture, where they can be used to lower the discharge of pollutants by improving the soil and/or their use in animal feed [74]. The cattle business is among the major causes of methane (CH_4) emissions. Excessive methane emissions directly affect climate change, as they aggravate greenhouse warming [91]. A previous study found that with a 2% increase in the alga species *Asparagopsis taxiformis* (Rhodophyta) in cattle feed, methane emissions decreased by 99% [74]. In a study focused on the agricultural cultivation of sugarcane, Singh et al. also found that *Kappaphycus* (Rhodophyta) extract as a fertilizer has the potential to lower CO_2 emissions. It further increased the economic value of the product [92].

5. Future perspectives

This emerging sector holds great potential—if well planned. Currently, the potential of seaweed aquaculture is not yet fully realized, and authors have pointed out "hurdles" that need to be overcome to do so [7, 21]. A lack of up-to-date guidelines and laws surrounding seaweed cultivation is present, and the provision of such is recommended. An example of this issue is represented by the currently poorly regulated genetic profile of seaweed cultivars 10, 11, 63, 69, 70. As the gene flow between cultivated and wild seaweeds cannot be prevented, well-informed regulations that determine the genetic diversity of cultivars should be implemented [7, 70]. However, currently, knowledge of this topic is limited [93]. On another note, spatial planning and the effects of implementing aquaculture have highly localized effects, and guidelines and recommendations for an approach to assessing and avoiding stressors could be advantageous [75, 94]. Such plans should also target currently under-addressed topics such as an excessive uptake of nutrients and altered flow regimes, which could negatively impact surrounding ecosystems. At this point in time, a lack of studies assessing such interactions is apparent [75].

Previous analyses have considered the potential of blue economy. To do so, the tools of quantification and monetizing the potential damages and benefits, such as economic values and prices paid by society and ecosystems, have been applied. Such benefits and externalities to the natural environment and human society can be measured, and a monetary value can be assigned according to cost-benefit analyses (CBA), which can serve as a tool in this process [70]. However, CBAs are not always objective, and results differ between studies [95]. Another approach to developing a critical step-by-step approach based in a SWOT (strength-weakness-opportunity-threat) analysis based on human risks, experienced environmental impacts and costs, constraints in economic growth, and feasibility of the aquaculture project [40].

There is a need for the promotion and growth of the sector globally and the overcoming of conflicts of interest presents. In particular, the promotion of partnerships between academics, companies, and communities to develop communication and technology for the purpose of creating long-term economic impacts is relevant. Moreover, there is a need for the expansion of the seaweed market. Government authorities and governmental capacities are key players in taking charge of expanding the sector [21].

6. Conclusions

Seaweed aquaculture offers a range of benefits to society, the environment, and the sustainable development of the economy. Under consideration of the overexploitation of wild seaweeds and an increasing interest in and application of seaweeds, it appears that aquaculture cultivation holds the future of commercial seaweed production to continue to meet the demand. However, at this point, only a fraction of the suitable areas for seaweed cultivation are utilized. Seaweed aquaculture has a large potential to grow, physically and as a market, if increased attention, research, and investment are devoted.

Seaweeds can be cultivated in the marine environment, nearshore, and offshore, as well as in saline land-based ponds. All methods are found to be viable, and none presents itself as highly favorable or disadvantageous. While the techniques of seaweed aquaculture systems, especially the integrated multi-trophic aquaculture systems (IMTAs), can be improved upon, they can support sustainability and a blue economy.

The potential relevance of seaweed aquaculture in the blue economy has become apparent. The sector can add to all components of the blue economy: economic growth, conservation, and social factors such as human wealth and health. Yet, several negative aspects, which especially affect the surrounding environments, need to be considered and overcome.

Seaweed aquaculture was also found to extend beyond the scope of the blue economy and reach into land and ocean-spanning climate change adaptation. Expanded seaweed aquaculture can provide services to reduce the effects of climate change and associated events, such as increased storm intensity and frequency, eutrophication, and ocean acidification. Furthermore, seaweed aquaculture can support the blue carbon framework due to its capacity for carbon sequestration. The use of seaweed aquaculture in counteracting the effects of climate change and correlated events is still in its infancy and should be further explored.

Acknowledgements

This research was co-financed by the project MENU—Marine Macroalgae: Alternative recipes for a daily nutritional diet (FA_05_2017_011), funded by the Blue Fund under Public Notice No. 5-Blue Biotechnology. João Cotas thanks the European Regional Development Fund through the Interreg Atlantic Area Program under the project NASPA (EAPA_451/2016). Diana Pacheco, thanks to PTDC/BIA-CBI/31144/2017—POCI-01 project-0145-FEDER-031144—MARINE INVADERS, co-financed by the ERDF through POCI (Operational Program Competitiveness and Internationalization) and by the Foundation for Science and Technology (FCT, IP). Alina Hillinger and Kay Ihle would like to thank IMBRSea—The International Master of Science in Marine Biological Resources. Ana M. M. Gonçalves acknowledges the University of Coimbra for the contract IT057-18-7253.

Conflict of interest

The authors declare no conflict of interest.

Author details

Madalena Mendes[1], Alina Hillinger[1,2], Kay Ihle[1,2], Miguel Cascais[1,3], Pedro Andrade[1], João Cotas[1], Diana Pacheco[1], João Carlos Marques[1], Leonel Pereira[1] and Ana Marta Mendes Gonçalves[1,4]*

1 CFE—Centre for Functional Ecology: Science for People and Planet, Marine Resources, Conservation and Technology—Marine Algae Lab, Department of Life Sciences, University of Coimbra, Coimbra, Portugal

2 IMBRSea, Ghent University, Ghent, Belgium

3 Higher Institute for Tourism and Hotel Management of Estoril, Estoril, Portugal

4 Department of Biology and CESAM, University of Aveiro, Aveiro, Portugal

*Address all correspondence to: amgoncalves@uc.pt

References

[1] Buchholz CM, Krause G, Buck BH. Seaweed and Man. Berlin, Heidelberg: Springer; 2012. pp. 471-493

[2] Pereira L. Edible Seaweeds of the World. Boca Raton, FL: CRC Press; 2016. Epub ahead of print 8 January 2016. DOI: 10.1201/b19970

[3] Yang LE, Lu QQ, Brodie J. A review of the bladed Bangiales (Rhodophyta) in China: History, culture and taxonomy. European Journal of Phycology. 2017;**52**:251-263

[4] Lobban CS, Paul JH, Paul JH. Seaweed Ecology and Physiology. Cambridge, UK: Cambridge University Press; 1994. DOI: 10.1017/CBO9780511626210

[5] Leandro A, Pacheco D, Cotas J, et al. Seaweed's bioactive candidate compounds to food industry and global food security. Life. 2020;**10**:140

[6] Stévant P, Rebours C, Chapman A. Seaweed aquaculture in Norway: Recent industrial developments and future perspectives. Aquaculture International. 2017;**25**:1373-1390

[7] Buschmann AH, Camus C, Infante J, et al. Seaweed production: Overview of the global state of exploitation, farming and emerging research activity. European Journal of Phycology. 2017;**52**:391-406

[8] Naylor RL, Hardy RW, Buschmann AH, et al. A 20-year retrospective review of global aquaculture. Nature. 2021;**591**:551-563

[9] Buschmann AH, Camus C. An introduction to farming and biomass utilisation of marine macroalgae. Phycologia. 2019;**58**:443-445

[10] Hafting JT, Craigie JS, Stengel DB, et al. Prospects and challenges for industrial production of seaweed bioactives. Journal of Phycology. 2015;**51**:821-837

[11] Ferdouse F, Holdt SLS, Murua P, et al. The global status of seaweed production, trade and utilization. In: Globefish Research Programme. Rome, Italy. p. 124

[12] Charrier B, Abreu MH, Araujo R, et al. Furthering knowledge of seaweed growth and development to facilitate sustainable aquaculture. New Phytologist. 2017;**216**:967-975

[13] FAO. The State of World Fisheries and Aquaculture 2020 Sustainability in Action. Rome: FAO; 2020. Epub ahead of print 2020. DOI: 10.4060/ca9229en

[14] Wood D, Capuzzo E, Kirby D, et al. UK macroalgae aquaculture: What are the key environmental and licensing considerations? Marine Policy. 2017;**83**:29-39

[15] Theuerkauf SJ, Morris JA, Waters TJ, et al. A global spatial analysis reveals where marine aquaculture can benefit nature and people. PLoS One. 2019;**14**:e0222282

[16] Thomsen M, Zhang X. Life cycle assessment of macroalgal ecoindustrial systems. In: Sustainable Seaweed Technologies. Elsevier; 2020. pp. 663-707

[17] The State of World Fisheries and Aquaculture 2022. The State of World Fisheries and Aquaculture 2022. Epub ahead of print 28 June 2022. DOI: 10.4060/CC0461EN

[18] The State of World Fisheries and Aquaculture - 2018 |Policy Support

and Governance| Food and Agriculture Organization of the United Nations. Available from: https://www.fao.org/policy-support/tools-and-publications/resources-details/en/c/1269332/ [Accessed: November 07, 2023]

[19] Kim JK, Yarish C, Hwang EK, et al. Seaweed aquaculture: Cultivation technologies, challenges and its ecosystem services. Algae. 2017;**32**:1-13

[20] Araújo R, Vázquez Calderón F, Sánchez López J, et al. Current status of the algae production industry in Europe: An emerging sector of the blue bioeconomy. Frontiers in Marine Science. 2021;**7**:1-24. Epub ahead of print 27 January 2021. DOI: 10.3389/fmars.2020.626389

[21] Rebours C, Marinho-Soriano E, Zertuche-González JA, et al. Seaweeds: An opportunity for wealth and sustainable livelihood for coastal communities. Journal of Applied Phycology. 2014;**26**:1939-1951

[22] García-Poza S, Leandro A, Cotas C, et al. The evolution road of seaweed aquaculture: Cultivation technologies and the industry 4.0. International Journal of Environmental Research and Public Health. 2020;**17**:1-42

[23] Sudhakar K, Mamat R, Samykano M, et al. An overview of marine macroalgae as bioresource. Renewable and Sustainable Energy Reviews. 2018;**91**:165-179

[24] Zheng Y, Jin R, Zhang X, et al. The considerable environmental benefits of seaweed aquaculture in China. Stochastic Environmental Research and Risk Assessment. 2019;**33**:1203-1221

[25] Hafting JT, Critchley AT, Cornish ML, et al. On-land cultivation of functional seaweed products for human usage. Journal of Applied Phycology. 2012;**24**:385-392

[26] Rosa J, Lemos MFL, Crespo D, et al. Integrated multitrophic aquaculture systems – Potential risks for food safety. Trends in Food Science and Technology. 2020;**96**:79-90

[27] Abreu MH, Pereira R, Yarish C, et al. IMTA with Gracilaria vermiculophylla: Productivity and nutrient removal performance of the seaweed in a land-based pilot scale system. Aquaculture. 2011;**312**:77-87

[28] Santelices B. A conceptual framework for marine agronomy. Hydrobiologia. 1999;**398-399**:15-23

[29] Nations U. Exploring the Potential of the Blue Economy. United Nations. Available from: https://www.un.org/en/desa/exploring-potential-blue-economy; [Accessed: November 06, 2023]

[30] The Potential of the Blue Economy - Increasing Long-term Benefits of the Sustainable Use of Marine Resources for SIDS and Coastal Least Developed Countries.:. Sustainable Development Knowledge Platform. Available from: https://sustainabledevelopment.un.org/index.php?page=view&type=400&nr=2446&menu=1515 [Accessed: November 06, 2023]

[31] The 2018 Annual Economic Report on EU Blue Economy - Publications Office of the EU. Available from: https://op.europa.eu/en/publication-detail/-/publication/79299d10-8a35-11e8-ac6a-01aa75ed71a1 [Accessed: November 06, 2023]

[32] Smith-Godfrey S. Defining the blue economy. Maritime Affairs. 2016;**12**:58-64

[33] Choudhary P, Subhash V, Khade M, et al. Empowering blue economy: From

underrated ecosystem to sustainable industry. Journal of Environmental Management. 2021;**291**:112697

[34] Hossain MS, Sharifuzzaman SM, Nobi MN, et al. Seaweeds farming for sustainable development goals and blue economy in Bangladesh. Marine Policy. 2021;**128**:1-9. Epub ahead of print 1 June 2021. DOI: 10.1016/j.marpol.2021.104469

[35] Keen MR, Schwarz AM, Wini-Simeon L. Towards defining the blue economy: Practical lessons from pacific ocean governance. Marine Policy. 2018;**88**:333-341

[36] Pauli GA. The Blue Economy: 10 Years, 100 Innovations, 100 Million Jobs. Boulder, Colorado; 2010

[37] Ahmed N, Thompson S. The blue dimensions of aquaculture: A global synthesis. Science of the Total Environment. 2019;**652**:851-861

[38] Hurtado AQ, Neish IC, Critchley AT. Phyconomy: The extensive cultivation of seaweeds, their sustainability and economic value, with particular reference to important lessons to be learned and transferred from the practice of eucheumatoid farming. Phycologia. 2019;**58**:472-483

[39] Froehlich HE, Afflerbach JC, Frazier M, et al. Blue growth potential to mitigate climate change through seaweed offsetting. Current Biology. 2019;**29**:3087-3093.e3

[40] Pratama MFA, Nurhayati A, Rizal A, et al. Strategy for the development of seaweed cultivation in Karawang Regency, West Java Province. Asian Journal of Fisheries and Aquatic Research. 2021;**11**:12-22

[41] FAO. World Fisheries and Aquaculture. Rome, Italy; 2018

[42] Campbell R, Hotchkiss S. Carrageenan industry market overview. In: Tropical Seaweed Farming Trends, Problems and Opportunities. Springer International Publishing; 2017. pp. 193-205

[43] Pereira L. Characterization of bioactive components in edible algae. Marine Drugs. 2020;**18**:65

[44] Future Market Insight Global & Consulting Pvt Ltd. Carrageenan Gum Market: Food & Beverages Application Segment Projected to Record Highest Revenue Growth Throughout the Forecast Period: Global Industry Analysis and Opportunity Assessment, 2016-2024. 2017. Available from: https://www.reportlinker.com/p04833615/Carrageenan-Gum-Market-Food-Beverages-Application-Segment-Projected-to-Record-Highest-Revenue-Growth-Throughout-the-Forecast-Period-Global-Industry-Analysis-and-Opportunity-Assessment.html [Accessed: May 24, 2021]

[45] Chung IK, Sondak CFA, Beardall J. The future of seaweed aquaculture in a rapidly changing world. European Journal of Phycology. 2017;**52**:495-505

[46] Sims R. Energy-smart food for people and climate. In: FAO Climate-Smart Knowledge Day Durban, 29 November, 2011. Rome, Italy; 2011

[47] Fernand F, Israel A, Skjermo J, et al. Offshore macroalgae biomass for bioenergy production: Environmental aspects, technological achievements and challenges. Renewable and Sustainable Energy Reviews. 2017;**75**:35-45

[48] Ashkenazi DY, Israel A, Abelson A. A novel two-stage seaweed integrated multi-trophic aquaculture. Reviews in Aquaculture. 2019;**11**:246-262

[49] Hasselström L, Visch W, Gröndahl F, et al. The impact of seaweed cultivation on ecosystem services - a case study from the west coast of Sweden. Marine Pollution Bulletin. 2018;**133**:53-64

[50] Michalak I. The application of seaweeds in environmental biotechnology. Advances in Botanical Research. 2020;**95**:85-111

[51] Besada V, Andrade JM, Schultze F, et al. Heavy metals in edible seaweeds commercialised for human consumption. Journal of Marine Systems. 2009;**75**:305-313

[52] Sadhukhan J, Gadkari S, Martinez-Hernandez E, et al. Novel macroalgae (seaweed) biorefinery systems for integrated chemical, protein, salt, nutrient and mineral extractions and environmental protection by green synthesis and life cycle sustainability assessments. Green Chemistry. 2019;**21**:2635-2655

[53] Kim JK, Kraemer GP, Yarish C. Use of sugar kelp aquaculture in Long Island Sound and the Bronx River Estuary for nutrient extraction. Marine Ecology Progress Series. 2015;**531**:155-166

[54] Kang YH, Kim S, Choi SK, et al. A comparison of the bioremediation potential of five seaweed species in an integrated fish-seaweed aquaculture system: Implication for a multi-species seaweed culture. Reviews in Aquaculture. 2021;**13**:353-364

[55] Krumhansl KA, Scheibling RE. Production and fate of kelp detritus. Marine Ecology Progress Series. 2012;**467**:281-302

[56] Kang YH, Shin JA, Kim MS, et al. A preliminary study of the bioremediation potential of Codium fragile applied to seaweed integrated multi-trophic aquaculture (IMTA) during the summer. Journal of Applied Phycology. 2008;**20**:183-190

[57] Holdt SL, Edwards MD. Cost-effective IMTA: A comparison of the production efficiencies of mussels and seaweed. Journal of Applied Phycology. 2014;**26**:933-945

[58] Silva DM, Valente LMP, Sousa-Pinto I, et al. Evaluation of IMTA-produced seaweeds (Gracilaria, Porphyra, and Ulva) as dietary ingredients in Nile tilapia, *Oreochromis niloticus* L., juveniles. Effects on growth performance and gut histology. Journal of Applied Phycology. 2015;**27**:1671-1680

[59] Santelices B, Doty MS. A review of Gracilaria farming. Aquaculture. 1989;**78**:95-133

[60] Buschmann AH, Westermeier R, Retamales CA. Cultivation of Gracilaria on the sea-bottom in southern Chile: A review. Journal of Applied Phycology. 1995;**7**:291-301

[61] Fleurence J, Morançais M, Dumay J, et al. What are the prospects for using seaweed in human nutrition and for marine animals raised through aquaculture? Trends in Food Science and Technology. 2012;**27**:57-61

[62] Emblemsvåg J, Kvadsheim NP, Halfdanarson J, et al. Strategic considerations for establishing a large-scale seaweed industry based on fish feed application: A Norwegian case study. Journal of Applied Phycology. 2020;**32**:4159-4169

[63] Boyd CE, McNevin AA, Clay J, et al. Certification issues for some common aquaculture species. Reviews in Fisheries Science. 2005;**13**:231-279

[64] Ferreira Barbosa D, Lúcia Pereira Dias T, de Faria Lopes S, et al. Community structure and functional traits of mollusks associated with coastal reef macroalgae in Northeastern Brazil. Marine Ecology. 2018;**40**:1-14. Epub ahead of print 1 October 2019. DOI: 10.1111/maec.12563

[65] Pfister CA, Altabet MA, Weigel BL. Kelp beds and their local effects on seawater chemistry, productivity, and microbial communities. Ecology. 2019;**100**:e02798

[66] Hughes AD, Corner RA, Cocchi M, et al. Beyond Fish Monoculture Developing Integrated Multi-Trophic Aquaculture in Europe. Florence, Italy; 2016

[67] Teagle H, Hawkins SJ, Moore PJ, et al. The role of kelp species as biogenic habitat formers in coastal marine ecosystems. Journal of Experimental Marine Biology and Ecology. 2017;**492**:81-98

[68] Vásquez JA, Zuñiga S, Tala F, et al. Economic valuation of kelp forests in northern Chile: Values of goods and services of the ecosystem. Journal of Applied Phycology. 2014;**26**:1081-1088

[69] Pacheco D, Araújo GS, Cotas J, et al. Invasive seaweeds in the Iberian Peninsula: A contribution for food supply. Marine Drugs. 2020;**18**:560

[70] Valero M, Guillemin M-L, Destombe C, et al. Perspectives on domestication research for sustainable seaweed aquaculture. Perspectives in Phycology. 2017;**4**:33-46

[71] Halling C, Wikström SA, Lilliesköld-Sjöö G, et al. Introduction of Asian strains and low genetic variation in farmed seaweeds: Indications for new management practices. Journal of Applied Phycology. 2013;**25**:89-95

[72] Krause-Jensen D, Marbà N, Sanz-Martin M, et al. Long photoperiods sustain high pH in arctic kelp forests. Science Advances. 2016;**2**:e1501938

[73] Hussain MG, Failler P, Al KA, et al. Major opportunities of blue economy development in Bangladesh. Journal of the Indian Ocean Region. 2018;**14**:88-99

[74] Riddick S, Snodgrass L, Swanson M. Innovation in the Blue Economy: Opportunities in Shipping, Aquaculture, and Offshore Wind for Climate Change Mitigation. Nicholas School of the Environment of Duke University; 2021

[75] Duarte CM, Wu J, Xiao X, et al. Can seaweed farming play a role in climate change mitigation and adaptation? Frontiers in Marine Science. 2017;**4**:100

[76] Moreira D, Pires JCM. Atmospheric CO_2 capture by algae: Negative carbon dioxide emission path. Bioresource Technology. 2016;**215**:371-379

[77] Xiao X, Agustí S, Yu Y, et al. Seaweed farms provide refugia from ocean acidification. Science of the Total Environment. 2021;**776**:145192

[78] Bari A. Our oceans and the blue economy: Opportunities and challenges. In: 10th International Conference on Marine Technology, MARTEC 2016. Elsevier Ltd; 2017. pp. 5-11

[79] Laurens LML, Nelson RS. Sustainable technologies for seaweed conversion to biofuels and bioproducts. In: Sustainable Seaweed Technologies. Elsevier; 2020. pp. 643-661

[80] Hill R, Bellgrove A, Macreadie PI, et al. Can macroalgae contribute to blue

carbon? An Australian perspective. Limnology and Oceanography. 2015;**60**:1689-1706

[81] Bayley DTI, Marengo I, Baker H, et al. Giant Kelp 'Blue Carbon' Storage and Sequestration Value in the Falkland Islands. Peterborough, UK; 2017

[82] Hurlimann S. How Kelp Naturally Combats Global Climate Change—Science in the News. 2019. Available from: https://sitn.hms.harvard.edu/flash/2019/how-kelp-naturally-combats-global-climate-change/ [Accessed: June 10, 2021]

[83] Ortega A, Geraldi NR, Alam I, et al. Important contribution of macroalgae to oceanic carbon sequestration. Nature Geoscience. 2019;**12**:748-754

[84] Queirós AM, Stephens N, Widdicombe S, et al. Connected macroalgal-sediment systems: Blue carbon and food webs in the deep coastal ocean. Ecological Monographs. 2019;**89**:1-21. Epub ahead of print 1 August 2019. DOI: 10.1002/ecm.1366

[85] Mashoreng S, Nafie L, Isyrini R. Cultivated seaweed carbon sequestration capacity Cultivated seaweed carbon sequestration capacity. IOP Conference Series: Earth and Environmental Science. 2019;**370**:1-9. Epub ahead of print 2019. DOI: 10.1088/1755-1315/370/1/012017

[86] Fakhraini MS, Wisnu W, Khathir R, et al. Carbon sequestration in macroalgae *Kappaphycus striatum* in seaweed aquaculture site, Alaang village, Alor Island, East Nusa Tenggara. IOP Conference Series: Earth and Environmental Science. 2019;**404**:1-8. Epub ahead of print 2020. DOI: 10.1088/1755-1315/404/1/012044

[87] Krause-Jensen D, Lavery P, Serrano O, et al. Sequestration of macroalgal carbon: The elephant in the blue carbon room. Biology Letters. 2018;**14**:20180236

[88] Keeling RF, Körtzinger A, Gruber N. Ocean deoxygenation in a warming world. Annual Review of Marine Science. 2010;**2**:199-229

[89] IPCC. Impacts, Adaption, and Vulnerability. Alistair Woodward; 2014

[90] Ross FWR, Boyd PW, Filbee-Dexter K, et al. Potential role of seaweeds in climate change mitigation. Science of the Total Environment. 2023;**885**:163699

[91] Ruppel CD, Kessler JD. The interaction of climate change and methane hydrates. Reviews of Geophysics. 2017;**55**:126-168

[92] Singh I, Anand KGV, Solomon S, et al. Can we not mitigate climate change using seaweed based biostimulant: A case study with sugarcane cultivation in India. Journal of Cleaner Production. 2018;**204**:992-1003

[93] Goecke F, Klemetsdal G, Ergon Å. Cultivar development of kelps for commercial cultivation—Past lessons and future prospects. Frontiers in Marine Science. 2020;**8**:1-17. Epub ahead of print 25 February 2020. DOI: 10.3389/fmars.2020.00110

[94] Eggertsen M, Halling C. Knowledge gaps and management recommendations for future paths of sustainable seaweed farming in the Western Indian Ocean. Ambio. 2021;**50**:60-73

[95] Harley CDG, Anderson KM, Demes KW, et al. EFfects of climate change on global seaweed communities. Journal of Phycology. 2012;**48**:1064-1078

Chapter 3

Deep Sea Biodiversity on the Continents, How It Happens?

Andrés Castrillón

Abstract

Recent studies in the ocean depths have discovered a large number of organisms and microorganisms that live in extreme environments of hydrothermal vents and cold seeps obtaining their energy through anaerobic oxidation of methane (AOM) process through a microbial consortium capable of reducing sulfate and oxidizing methane simultaneously. On the surface, the fossil record of this type of chemosynthetic community has made it possible to establish the link between the tectonic plate dynamics with the origin of mineral deposits or sediments formed on the oceanic ridges and attached to the continents. The foregoing could be supported by the ^{13}C and ^{18}C stable isotopes analysis that has been carried out in the fossil record of rocks and sediments attached to the continents and the study of stable isotopes that is currently carried out in submersibles at mid-ocean ridges around the world. The relationship between current values and the fossil record is key to understanding, among others, the methane contribution to the environment and its variation throughout time. Cerro Matoso is a recent case that permitted linked the recent hydrothermal activity in the mid-ocean ridge, with those of the ancient ones.

Keywords: vent-cold seep, mid-ocean ridges, tectonic plates, vent communities, anaerobic methane oxidation

1. Introduction

The evolution of the oceans is linked to the origin and geological evolution of the continents. The Plate Tectonics theory, which led to the understanding of the continental margin's evolution and the constant generation of oceanic crust of slow and ultra-slow spreading along different oceanic ridges, has been supported by the discovery and identification onshore of mantle and marine sedimentary rocks attached to the continents and exposed in impressive outcrops that show the impressive terrestrial dynamics.

As an example, one of the most important events in the geological record is related to the early Cambrian biological evolution, time during which the intracontinental extension was intensively dealing with hydrothermal activity in submarine settings [1] and black shale with multi-metal beds was formed and deposited around the world, are represented today with oceanic sediments attached to the continents and exposed on surfaces mainly in Asia and Europe. A series of hydrothermal vent community discovered at the Cambrian base in black shales indicated that the hydrothermal vent existed in the extensive ecological community in the early Cambrian [2]. A lot of these early Cambrian vent communities are similar to the modern Eastern Pacific vent fauna. Therefore, the hydrothermal vent communities make up the main ecological community in the early Cambrian [2].

Recently, the studies of different hydrothermal systems that exist in the oceanic ridges (e.g., in Refs. [3–5]), enlightened the immense biodiversity around these environments where microorganisms live from the biosynthetic relationships that are generated with the organisms that hosted them. Studies of stable isotopes of ^{13}C and ^{18}O in the calcareous sediments at these mid-oceanic ridges and about the hydrothermal fluids that they host indicate that much of the carbon forming carbonates comes from the mantle and is generated from Fyscher–Tropsch reactions (FTT) ($4H_2 + CO_2 -> CH_4 + 2H_2O$), which is an abiotic process (e.g., Refs. [6, 7]).

Hydrothermal activity is a phenomenon that takes place all over the planet and refers to the circulation of fluids within the mantle and the hot Earth's crust. Manifestations of these processes that take place in the crust can be observed at the bottom of the sea (black or white smokers). The diversity of communities associated with hydrothermal vents in the mid-Atlantic oceanic ridge–MAR, has been described [8–10] and are considered exceptional in their biomass, distribution, and composition [10, 11].

2. Hydrothermal vent fauna discovery

In 1977 the famous deep-sea submarine ALVIN, on the Ocean Pacific Ridge near the Galápagos Islands revealed incredible benthic communities of giant clams, tube worms, and microbial films that were associated with warm benthic fluids at temperatures over 370°C and emanation from black smokers [12]. That's how was discovered rich and unusual colonies that owed their existence to chemosynthetic microorganisms fed by magmatic degassing and water-rock reactions associated with the convection of warm waters along the Galápagos ridge [13], which confirmed the existence of active submarine hydrothermal systems windows and black smoker [14].

Constant investigations of the seabed have revealed that deep-sea hydrothermal windows harbor the most primitive sulfate-reducing, thermophilic, and chemosynthetic organisms that are implicated not only because of their biological adaptation in dark ocean-floor environments but also because they are related to the ancient forms of life on the planet [15–18].

The deep-sea hydrothermal systems expel large amounts of CO_2 from the earth's mantle, the isotopic carbon footprint of the mantle is recorded in the sediments and calcareous shells of the organisms that live in these environments. Recent studies of the stable isotopes of ^{13}C and ^{18}C carried out in the waters and sediments in these settings have allowed us to recognize the geological record of sediments attached to the continents. The hydrothermal sediments hosted by the ultramafic rocks of Cerro Matoso, in Colombia northwestern, are an example of these processes and how the earth's dynamic systems seem to be similar today to that from the Jurassic or earlier in the oceanic ridges where rocks and sediments traveling through the ocean crust are in some occasions, attached to the continents in active tectonic plates regimes. According to Campbell [19], there are now at least 59 regional groupings of early Archean to Pleistocene age ore deposits and marine sedimentary sequences worldwide related to submarine origin.

Some examples are found on the Eurasian plates and the old continent. Fossils of early Cambrian hydrothermal vent communities in Guizhou Province, China, that might be the oldest hydrothermal vent community ever reported [2] or the hydrothermal vent worms from Cretaceous sulfide ore of the Samal Ophiolite, Oman [20] or the Silurian hydrothermal vent community from the southern Urals, Russia and Late Cretaceous hydrothermal vent communities from the Troodos ophiolite, Cyprus [21, 22] are indicative that the oceans are linked with the continents. Other examples are located in the American continent, where oceanic crust generated in the mid-oceanic Pacific ridge, is attached to the western part of the American continent along the oceanic trenches, when segments of oceanic crust instead of sinking under the continent, manage to ascend and attach themselves to it.

Now it is known that hydrocarbon vents and seeps are common at continental margins and oceanic spreading ridges worldwide, exuding fluids rich in CH_4 and H_2S, and teeming with life based on chemosynthesis. These settings have been implicated as the crucibles for life's origin, and as locals for methane release to the atmosphere from hydrate destabilization during the past climate change [19] According to Campbell, hydrothermal vents and seeps are necessary components of climate models, because cycling of methane-derived carbon from the lithosphere to the hydrosphere and atmosphere includes emissions from the 75,000 km-long oceanic ridge system, and from seepages or gas hydrates released around the world's continental margins (e.g., Refs. [23, 24]).

Extraordinarily around the world, in some places, these tectonic processes of collision between tectonic plates, allowed the rocks to preserve part of their geological record formed in the ridges, perhaps the deformation front or contact zone between the plates was far from these rocks enabling the hydrothermal or seep fauna preservation, denoting that the detachment and deformation zones did not affect them (**Figure 1**).

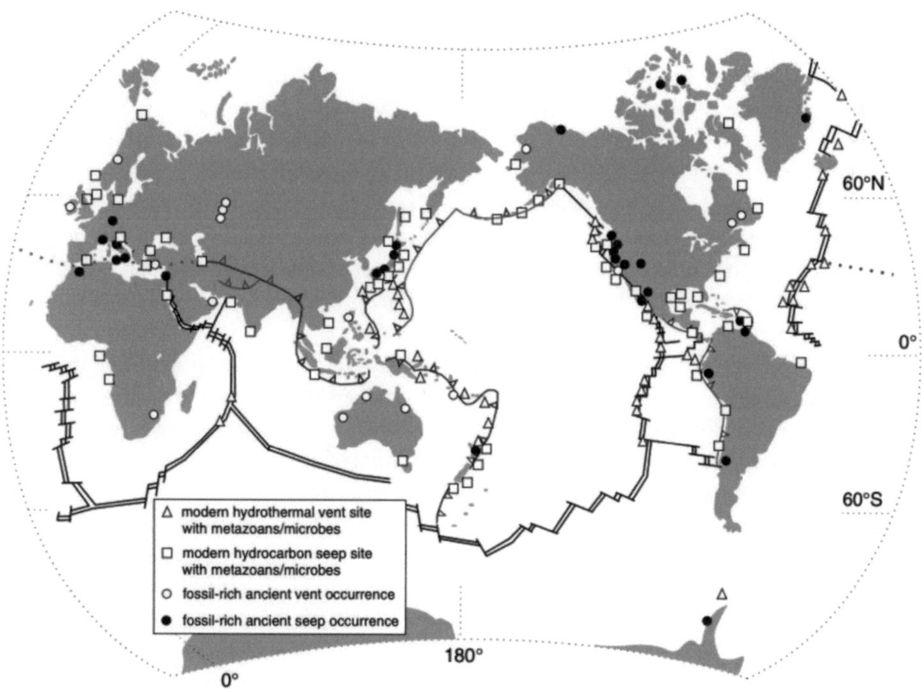

Figure 1.
Distribution map of chemosynthesis-based settings, illustrating those Archean to recent hydrothermal vents and hydrocarbon seeps with associated metazoan and/or microbial signatures [19].

3. Vents

The first of these abiotic associations were recognized in the Cretaceous massive sulfide deposits in the sulfide ore of the Samal Ophiolites, Oman [20] since then they have been known to exist throughout the Phanerozoic [19]. One of the most recent microbial ecosystems was discovered along the Mid-Atlantic Ridge and transform faults where H_2, CH_4, and other elements are emitted from the deep sea in serpentinite outcrops [25]. Many vent communities are reported and studied today on the oceans deeps, although many of them have their ancestors in fossils of benthic communities, some findings will correspond to new species. The interaction between carbonates and tube worms is not yet known, it is possible that worms grow concomitantly, creating void and channels that favor the rise of hydrothermal fluids [26].

Additionally, the findings of fossilized hydrothermal vents communities allowed us to turn our regard to the oceans, and recognize their immense activity and diversity of ecosystems, so we can relate the fossil fauna of Cenozoic and some Mesozoic hydrocarbon seeps with those that exist today, which are dominated by mollusks; in contrast to the fauna that dominates seeps in most Mesozoic and Paleozoic deposits, which is dominated by brachiopods [27].

Currently, it has been possible to identify that hydrothermal benthic communities vary from the Mid-Pacific Oceanic Ridge to the Mid-Atlantic Ridge; the former is dominated in terms of biomass by a group of endemic taxa that form endosymbiotic relationships with chemosynthetic bacteria [22]. This group includes the vestimentiferan tube worm (=obturate pogonophoran), gastropods, mytilids (*Bathymodiolus* species), and vesicomid bivalves (*Calyptogena* species). Atlantic benthic communities are dominated by chemosymbiotic *Bathymodiolus* species and several species of Bresiliid shrimp (e.g., *Rimicaris exoculata*, in Ref. [28]), which can feed on exosymbiotic bacteria growing on their bodies [22]. Chemo-symbiotic taxa have important physiological and morphological adaptations to live in their habitats, an example being the loss or reduction of the alimentary tract in adults [22] or slower growth in related taxa at ocean depths [29, 30]. The habitat complexity leads to considerable amounts of heterogeneity in animal communities [30, 31]. An important aspect of modern benthic communities is the large number of endemic taxa that they possess so it is suggested that some endemic species represent Mesozoic and even Paleozoic remnants and that places of hydrothermal emanations acted as refuges to prevent their extinction [22].

Although it is unlikely that ancient hydrothermal deposits in the Archean oceanic crust are well preserved and both fossils and microfossils are easily identifiable, in Colombia, specifically at the Cerro Matoso nickel mine, a community of deep-sea benthic organisms is fossilized. This community hosted in a serpentinized rock attached to the South American continent during the Cretaceous, is forming bioherms where extremophile organisms could apparently survive under the influence of hydrothermal vents [32]. The history of the evolution of the biological communities associated with hydrothermal vents is little known, only 19 occurrences of fossils hosted in massive sulfide deposits have been found, all of them Phanerozoic in age: three from the Silurian in the benthic hydrothermal community south of the Urals Russian [21] six Devonian, one Jurassic, seven Cretaceous, one Late Cretaceous in the Trodoos Ophiolites on Cyprus [21] and one Eocene [22]. Few works have presented evidence of microbiological communities or microbial alterations in ophiolites and Archean greenstone belts [33].

4. Cold seeps

A second type of chemosynthetic oasis, called cold seeps, was discovered during dives in the Gulf of Mexico and Pacific subduction zones depths [12] and related to hydrocarbon emissions, mainly methane but also ethane, propane and even oil [34] that seep through the sediment through cracks or fractures at flow velocities of a few tens of centimeters to a few meters per year [35], have been related to advection processes by the subduction of tectonic plates, due to processes of compaction and dehydration of hemipelagic sediments, a significant volume of water lost by sediments is expelled together with fluid mud and is channeled through deep faults until it is expelled on the ocean floor, where they manifest as mounds of mud with emanation windows often populated with benthic communities and covered by authigenic carbonates [26].

These cold seeps are associated with the dehydration of active and passive continental margins worldwide that generate favorable conditions for the precipitation of authigenic carbonates [26], which is mainly driven by the anaerobic oxidation of methane (AOM) through a microbial consortium capable of reducing sulfate and oxidizing methane simultaneously (e.g., in [36–41]).

Anaerobic microorganisms that mediate methane oxidation in AOM environments with sulfate, according to the equation $CH_4 + SO_4{}^{2-} \rightarrow HCO_{3-} + HS^- + H_2O$, carry out the dominant process in cold seep ecosystems and are one of the main methane consumers worldwide. Methane oxidation end products, such as bicarbonates and sulfides, are released into the surrounding rock or sediment and lead to the formation of authigenic carbonates that precipitate in sediments near the ocean floor at the bottom of a zone called the methane–sulfate interface (SMI) [26], which varies in depth depending on the rate of methane supply from the bottom and sulfate from above.

The precipitation of carbonates (high-magnesium calcite –HMC–, aragonite, or dolomite) close to the seafloor at cold seeps also results from anaerobic oxidation of methane (AOM), in which methane is oxidized using sulfate as an electron acceptor releasing bicarbonate and sulfide into pore-water [42].

Because oxygen is easily depleted below the ocean floor, subsurface anaerobes known as methanotrophs, hydrocarbon degraders, and sulfate-reducing bacteria, are the key functional groups in cold seep ecosystems [12]. The sulfur produced by these microorganisms can feed an entire ecosystem, microbial colonies, or faunas that can be found between 100 cm^2 and hundreds of square kilometers in diameter.

The relatively recent discovery of fossil fauna associated with seepage of hydrocarbons on the continents, both in Cenozoic rocks (e.g., in the western United States, Japan, and Italy) and Mesozoic (e.g., in the western United States, Japan, and France) and even Paleozoic records (e.g., in Germany, Morocco, and Mexico), have allowed us to understand the origin and types of mineralization of various formations. Additionally, the presence of ancient fauna of marine origin on the continents gives us signs and evidence of the Earth's dynamics.

Taxonomic groups have been reported from the Cretaceous to recent deposits associated with hydrothermal vents or cold seeps [21, 22, 43]. These groups include a variety of tubeworms, *Bathymodiolus* bivalves, and some bryozoans, among others. Members of the Bathymodiolina family occupy modern hydrocarbon seep and hydrothermal window environments [44]. Brachiopods are common in Paleozoic and Mesozoic seeps, infrequent in Cenozoic seeps, and almost absent in modern seeps and seeps [19]. Polychaetes of the order Eucidina are recognized as important members of Paleozoic marine benthic communities [45]. Modern seep communities

Tube Diameter (mm)	Age	Reference
0.86 to 8.75	Devonian,	[27]
30	Devonian,	[46]
15 to 30	Carboniferous,	[47]
10	Jurassic? Toarcian	[48]
0.2 to 0.5 and 2 to 5	Late Jurassic to Late Cretaceous	[49]
1 to 4	Cretaceous, Coniacian to Santonian	[50]
3 to 25	Paleogene, Paleocene	[51]
more of 5	Paleogene, Eocene	[52]
6	Paleogene, Eocene	[53]
more of 2.9	Paleogene, Oligocene	[54]
more of 7	Paleogene, Oligocene	[55]
1 to 3	Neogene, Miocene	[56]
0.7 to 1	Neogene, Miocene	[57]

Table 1.
Tubeworms occurrences in ancient hydrocarbon-seep deposits [27].

are dominated by sibloginid tubeworms (Vestimentifera and Frenulata) as well as bivalves (Vesicomyidae, *Bathymodiolus*, Lucinidae, Thyasiridae and Solemyidae), all relying on endosymbiosis. Calcareous tubular fossils are uncommon in Phanerozoic seep deposits (**Table 1**) but have been tentatively identified as vestimentiferous.

Current knowledge of the microbial diversity around the vent-seep reveals that ε-proteobacteria is perhaps the key to the carbon, nitrogen, and sulfur cycle and have an important role in the symbiotic association with benthic metazoans [19]. Given that the Earth in its beginnings had a warmer environment, without atmospheric oxygen, rich in components such as CO_2 and H_2, perhaps life could flourish in warm environments very similar to today's hydrothermal systems, they may also represent analogies with ancestral niches of life and provide the basis for the interpretation of biomarkers of ancient hydrothermal systems [58]. If hydrothermal activity has continued through geological time, thermophilic communities have probably evolved to maintain themselves in these environments, leaving biomarkers in the rock and fossil record that, together with isotopes and minerals preserved in modern or ancient deposits, allow comparison of current organisms with ancient microbial communities and paleoenvironmental interpretations [58].

Cold seep sediments host a large proportion of sulfate-reducing bacterium and methane-oxidizing Archaea, mainly methanotrophic and thiotrophic species such as Euryarchaeota and Crenarchaeota, the former living in chemical symbiosis in clams and mussels, while the thiotrophic species live in tube worms [34]. Carbon isotope analyzes in seep ecosystems also have shown a close link between methane, archaeal methanotrophils, and their sulfate-reducing partners, precipitated authigenic carbonates, and higher trophic levels in the food web [59]. Thus, like hot seeps, cold seeps support an enormous free symbiotic microbial biomass that lives and thrives on the oxidation of methane, higher hydrocarbons, and sulfide [34, 60] that are hosted by the more fascinating organisms, such as tube worms, clams, crabs, and mussels, observed in these environments.

The nature of these ecosystems based on the microbial chemical synthesis of H_2S seeps at high temperatures along mid-ocean ridges [61] presents an excellent analogy to anomalous fossil associations in the geological record of deep-sea marine environments commonly mineralized in base metals [62]. Ancient vent and seep deposits are also increasingly recognized and occur in various sizes, lithologies, biotic compositions, geotectonic settings, and ages.

5. Oceanic crust attached to the continents

Around de world, there are various ophiolites exposed on the surface that have been targets of resources investigation because of their mineral content, but also because some of them contain fossil evidence of ancient vent communities. Examples are in the Northern Apennines where ophiolites are considered to have formed at various stages in the development of a mid-ocean ridge system in the Ligurian Tethys as the result of the divergence of the European and Adriatic plates in the Middle Jurassic (e.g., in [63]). In Morocco, the Aït Ahmane ultramafic unit of the ca. 760 Ma Bou Azzer ophiolite (Morocco) hosts a fossil black smoker-type hydrothermal system [64].

Ophiolites are formed because of the mid-ocean ridge generation that is driven by magmatic and tectonic processes, which result from seafloor spreading. Tectonically dominated spreading and the formation of detachment faults cause mantle rocks to be exposed to seawater (e.g., [65–68]). In the case of slow-spreading ridges (full spreading rate $< \sim 5$ cm/year), spreading is accommodated by the formation of a conjugated faults network that, in many places in the Atlantic Ocean, lift mantle peridotites up to the ocean floor (e.g., [6]). This phenomenon is referred to as tectonic accretion in opposition to magmatic accretion in the case of basaltic systems [6]. Ultramafic rocks may form domes associated to transform faults perpendicular to oceanic spreading ridges, where the detachment fault arrays produced thereby drag lower crustal magmatic chambers and triggered the emplacement of shallow-depth gabbro intrusions [67, 69–71]. This creates favorable conditions for the hydrothermal circulation systems in the serpentinite slab brought to the seafloor surface [69].

Some examples of how rocks from the mantle are exposed in mid-oceanic ridges comprise basaltic and gabbroic basement hosting high temperature (350–400 C) hydrothermal systems (e.g., in [70, 71]), however oceanic serpentinites exposed on the ocean floor in the MARK area [72] and at the 15,200 N fracture zone along the Mid-Atlantic Ridge [73] or the peridotite-hosted system at the Atlantis Massif along the Mid-Atlantic Ridge [74] has shown that hydrothermal systems can vary from cold to high temperatures.

It is proposed that the breaking-up of the oceanic plates in heterogeneous crust fragments does not produce vertical planes, but instead deep and slightly inclined faults appear to control most of the slow movement of the plate, thus exhuming and exposing lower crust and mantle rocks [75]. Some of these rocks are exposed as oceanic core complexes (OCC), which are deep sections of the oceanic lithosphere exhumed to the seafloor by long-lived detachment faults formed along with slow and ultra-slow spreading centers at different scales [67, 76–78]. The OCC are formed perpendicular to the ridges near the end of the slow spreading segments, although their evolution could suggest that fracture zones do not become locked immediately on transform-to-fracture transition [67]. The domes can exceed 100 m of relief high, several hundred meters in width, and up to kilometers in length [75] conditions that could be compared with the case of Cerro Matoso peridotites located northwest of Colombia in the Caribbean region.

Precisely, the most recently studied example that allows establishing the relationships between the oceans and continents, their biodiversity, and the atmosphere, is in Cerro Matoso, where peridotite as part of ophiolitic segments was attached to Western Colombia controlled by complex geometries at the time of the collision between the Caribbean-, Nazca-, Cocos- and South American Plates.

The oceanic crust that builds today in the Central Caribbean region is Cerro Matoso, which was formed during the Cretaceous, about 89 Ma ago [79]. The thick oceanic crust of the Caribbean plate appears to be the tectonized remnant of an eastern Pacific oceanic plateau that has been inserted between North and South America. The emplacement of the plateau into its present position has resulted in the abduction and exposure of its margins, providing an opportunity to study the age relations, internal structure, and compositional features of the plateau.

The synchronicity of ages across the region is consistent with a flood basalt origin for the bulk of the Caribbean plateau (i.e., large volume, rapidly erupted, and regionally extensive volcanism in the Pacific Ocean), far west of its present position [80, 81]. As the American continent was moving to the west during the Late Cretaceous and the oceanic plate in turn was moving eastwards, oceanic segments were accreted in the western part of the continent [82]. Aspects of the collision and subduction or accretion in which the geometry of the process prevents an ophiolite segment from entering the subduction zone and instead preserves it against continental crust at shallow levels were discussed by Ref. [82].

In the Southern Caribbean, oblique convergence also resulted in the accretion of peridotites with high-pressure associations that may have started from the Eocene (55–50 Ma) and completed in the Oligocene or even in the early Miocene (30–15 Ma ago) [82]. These high-pressure associations correspond with bodies accreted to the western edge of the South American Plate during the Eocene [83].

6. Stable isotopes

Carbon and oxygen stable isotopes in vents and seep-carbonates are biogeochemical archives of past fluid activity and composition, and their measured values in the fossil content or substrate sign the geological history of these deposits (e.g., in Ref. [19]). The classic, isotopically depleted, carbonate-carbon signals recorded in some seep deposits and hydrothermal vents, were derived from microbially mediated, anaerobic oxidation of methane (AOM) in the sulfate reduction zone (e.g., in Refs. [19, 26]). However, variation in both carbonate-carbon and oxygen values is dictated by several processes, including kinetic isotope fractionation, hydrate dissociation, variable mixing of carbon sources, tapping of deep burial fluids (via faults or erosion), clay mineral dehydration, and diagenesis (e.g., in Refs. [19, 26]). Biogeochemical conditions in cold seep systems as well as in deep-sea hydrothermal systems depend on the flux of methane, availability of sulfate, temperature, and redox conditions. These conditions promote the development of chemosynthetic communities mainly dependent on methane and/or sulfur, driving microbial reactions that alter sulfur and carbon pools and whose pathways involve the formation of authigenic minerals (e.g., in Ref. [4, 5]).

Oxygen isotope analyses of marine carbonate fossils could substantiate the modeled paleoclimatic evolution by providing information on oceanic paleoclimatic temperatures [84], while carbon isotopes can be used to document changes in $\delta^{13}C$ values of ancient oceanic dissolved inorganic carbon (DIC). In this sense, stable isotope and element composition of the sediments and rocks are used to reconstruct the origin and nature of the fluids from which they formed (e.g., in Ref. [19]). However, isotope ratios of coeval carbonate shells of various organisms may differ from one another and only a study of a well-preserved fossil shell assembly could provide clear information on temperature variations within the water column. Nonetheless, bulk sample isotope analysis also gives good information about the conditions under there were formed.

Different approaches can be used to characterize the nature, source, and evolution of fluids benthic by examining the mineralogy, petrography, and isotopic characteristics of the carbonates [26]. Considering the link between hydrothermal vents, hydrocarbon-rich cold seeps, gas hydrate formation, and sub-seafloor hydrocarbon reservoirs (e.g., in Ref. [85]), the understanding of the occurrence of authigenic mineralization is crucial for linking the geological record of the rocks.

As the hydrothermal vents hosted by ultramafic rocks have been recently studied (e.g., in Refs. [4, 25]), the source of fluid hosted by them has been related to the serpentinization process that suffers peridotites when they are under intense activity provided by the heat of intrusion of mafic magmas or by the cooling of the ultramafic lithosphere, the later results in a fluid circulation that is driven by convection-dissipated heat from the bottom and exothermic chemical reactions between the circulating fluids and host [73]. Serpentinite-derived fluids in both deep-sea and ophiolite environments, where methane has been measured, present negative $\delta^{13}C$ values, including −11.9‰ at Lost City, −10.3‰ at Logatchev, −16.7‰ at Rainbow, −18‰ at Elba, and − 7.7‰ in the Zambales ophiolite [86–92]. Hydrothermal systems hosted in ultramafic rocks are present in Logatchev, Rainbow, and Lost City Hydrothermal Field-LCHF [3]. High-temperature black smokers generated by the cooling of different mixtures of gabbro and deep ultramafic materials are usual in these environments [3].

Some hydrothermal fields with associated calcareous sediments (Lost City and Rainbow) have been especially well studied [4, 88, 93–95]. In the "Clamstone" site on Rainbow Hill, the carbon isotopic composition of Thyaira aff. Southern shells showed signatures of more depleted ^{13}C ($\delta^{13}C = -7.69 \pm 1.60$‰) than expected, which is interpreted by the authors as a contribution of oxidized methane in the pore-water filling of the burrows formed by bivalves [96]. The formation of these sedimentary structures is analogous to fluid-induced chimney formation on modern vents. A record of these environmental conditions and isotopic signatures is in Cerro Matoso.

7. Cerro Matoso case

Cerro Matoso is an open-Pit mine, which is mined, molten, and produced a Fe-Ni allay from the laterites formed as a product of the peridotites weathering. Locally, on the peridotites, a community of benthic organisms that originated in the ocean depths is fossilized. The community is hosted in sediments overlaying the peridotites and is forming bioherms, where extremophile organisms could survive under the influence of hydrothermal emanation vents. Two types of sediments identified as green fossiliferous claystone and black mudstone contain evidence of having been generated under the influence of hydrothermal activity at intermediate temperatures [97]. The ^{18}O and ^{13}C isotopic reactions of both types of sediments (i.e., claystone and mudstone), analyzed in bulk sediments are consistent with those observed in serpentinization settings of ultramafic rocks (e.g., in Refs. [62, 92, 98]).

The mineral precipitation temperatures of the green fossiliferous claystone, obtained through the [99] relationship of the $\delta^{18}O$ values are in the range of 30 to 80°C, and although the lowest values could be interpreted as surface temperatures considering meteoric water, which is supported by the relatively negative $\delta^{13}C$ values, the $\delta^{18}O$ values indicating temperatures close to 80°C are signing a hydrothermal input. The precipitation temperatures of the black mudstone in a wider range of $\delta^{18}O$ values (15.9 to 29.8‰ V-SMOW) are better explained by the influence of low temperature hydrothermal/diagenetic system, however, the values of $\delta^{18}O$ (25.1 to 29.8‰ SMOW) reported in the intraclast within black mudstone are signing higher temperatures that only can be justified if fluids are enriched in ^{18}O (magmatic, metamorphic or diagenetic water). The corresponding lower $\delta^{13}C$ values (in intracalst, −27‰ V-PDB) could only be explained by these processes or by the oxidation of methane related to these systems. Contrary to the Cerro Matoso evidence, other smectites produced by ultramafic weathering [100] with $\delta^{18}O$ values (20.3 to 24.3‰ SMOW), in the nickel smectites in Murrin Murrin, Western Australia, represent isotopic $\delta^{18}O$ values that correlate with water-rock interactions at low temperatures due to the contact of meteoric fluids during mineral leaching processes in which laterites are formed [101].

Marine/diagenetic waters with relatively negative values of Dissolved Inorganic Carbon (DIC) are formed by the biological degradation of organic matter in closed systems [102], however, there is no obvious source of significant organic material (e.g., carbon-rich shales) in the sedimentary succession hosting the Cerro Matoso peridotites, so the production of abiogenic methane by serpentinization of peridotite looks like the more possible light $\delta^{13}C$ source. Methane in modern serpentinite-derived fluids in both the deep-sea and in ophiolite environments has negative $\delta^{13}C$ values (e.g., −11.9‰ at Lost City, −10.3‰ at Logatchev, −16.7‰ at Rainbow, −18‰ at Elba and – 7.7‰ in the Zambales ophiolite) similar to those reported in Cerro Matoso. Some of this methane will be oxidized close to the seafloor to produce ^{13}C-depleted DIC, and the high pH and Ca concentrations in the serpentinite-derived fluids promote carbonate and mineral precipitation on mixing with seawater [103, 104].

The fluids generated during the peridotite serpentinization favored the calcareous sedimentary succession formations found on top of the crystalline rocks of Cerro Matoso when the mantle rocks were exposed in an ancient mid-oceanic ridge. The $\delta^{13}C$ values between −27.1 and – 17.1‰ (i.e., calcareous intraclast within the black mudstone), indicating a low contribution of marine DIC surely associated with anaerobic oxidation of methane process–AOM, is interpreted as a contribution of oxidized methane in the pore-water filling the burrows formed by the bivalves (e.g., Ref. [90]). The Cerro Matoso ^{13}C-isotope data are in the range of serpentinite samples of six different regions in the Mid-Atlantic Ridges (MAR) and yielded $\delta^{13}C$ values ranging from −29 to −4‰ [5], which sign methane sources. Owing to the very low concentration of inorganic carbon in serpentinization fluids, the most feasible origin for this ^{13}C depletion is the oxidation of methane [105]. The rare rock fragments characterized by altered ultramafic lithoclasts (serpentinites) and claystone are in part similar to those described in Saldanha sediments, a hydrothermal site of the MAR partly supported by serpentinization processes [106].

Additionally, the $\delta^{13}C$ values in Cerro Matoso agree in part with the values of $\delta^{13}C$ of CH_4 and CO_2 of high- and low-temperature hydrothermal fluids, at 9° 50'N East Pacific Rise, with an average value of $\delta^{13}C$ of −30.2 ± 2.7‰ and – 4.5 ± 0.53‰, respectively, which are somewhat more diminished than the average values of $\delta^{13}C$ in high-temperature vents, −20.1 ± 1.2‰ of CH_4 and – 4.08 ± 0.16‰ of CO_2 [89]. These values are among the lightest measured in hydrothermal vents

[30, 87, 107, 108]. In general, carbon dioxide derived from magmatic degassing, which retains values of $\delta^{13}C$, mainly between −4 and − 8‰ [109], and carbonates in the mantle, between −5 and − 7‰ [110, 111], will be related only to the most negative $\delta^{13}C$ isotopic signature of the marine sediments in Cerro Matoso because in general, all samples present a wide range of $\delta^{13}C$. The more negative $\delta^{13}C$ values point out the presence of methane oxidation events, favoring the growth of carbonates in the sediments overlaying the ultramafic rocks. Similar conditions have been reported around deep marine hydrothermal vents [5, 25, 92].

The initial isotopic carbon ratio of the carbon dioxide discharged from geothermal and ultramafic systems in oceanic ridges, which ranges from −8‰ to −3‰, plus the maximum fractionation factor resulting from the oxidation of anaerobic or aerobic methane (1.039) at hydrothermal environments around ultramafic rocks [89, 112] favoring depletion by at least −13‰ of the original isotope signature, giving $\delta^{13}C$ values around −21‰ that are consistent with the negative isotopic signature of the minerals forming marine claystone and mudstone in Cerro Matoso. These values may also record anaerobic oxidation of methane or bacterial oxidation/sulfate-reduction [102].

Evidence that methane emissions from mid-ocean ridges have persisted since their origin, and that their influence can be analyzed from the fossil record in rocks originating from mid-ocean ridges, but that today are attached to the continents, may be verified from the study of the isotopic fractionation models for ^{13}C and ^{18}O (e.g., in Ref. [113]). These models allow us to compare the expected behavior of the fraction that reacts in the process of gain and loss of heavy and light isotopes, in an open system in the presence of methane and sulfur, such as the one that occurs in the mid-oceanic ridges (e.g., Ciudad Lost, Logatchev). In the case of Cerro Matoso, a hydrothermal system has been considered, similar to the recently studied Lost City hydrothermal field, on the Mid-Atlantic oceanic ridge, where ultramafic rocks are exposed on the seabed and host chemosynthetic communities and hydrothermal vents that manage to resemble what was observed in the outcrops of the open-pit nickel mine of Cerro Matoso. However, the results reflect the existing pre-Cretaceous conditions in the ancient mid-oceanic Pacific ridge, where the Cerro Matoso peridotites must have been exhumed hosting hydrothermal systems that allowed the vent communities to be hosted on the rocks and marine sediments.

For the isotopic modeling of ^{13}C in the Cerro Matoso green fossiliferous sediments (**Table 2**), an environment of warm waters in the ocean depths was considered, in which H_2S forms a group of isotopes together with CH_4, OH^-, HCO_3^- and Fe, among other elements, and where $CaCO_3$ and CH_4 are products of the isotopic exchange during the process of loss and gain of electrons. In this environment, the fractionation factor resulting from anaerobic or aerobic oxidation of methane can be as high as 1.039 [89, 112, 114]. Based on this value, and assuming that there was an anaerobic methane oxidation process, the ^{13}C fractionation model considers the maximum value of isotopic fractionation Δ = 39‰, for product 1 or $CaCO_3$. The residual substrate in these reactions, in our case, the product 2 or (CH_4), assumes an isotopic fractionation Δ = 12.3‰, values that are in agreement with the laboratory results of $\delta^{13}C$, and represent the highest isotopic depletion in the seawater. Maintaining the mass balance, the model made (**Figure 2**) shows the behavior of the expected isotopic variations (solid lines red and blue), for products 1 and 2, and the variations that would occur in the internal pool of isotopes (dotted line in red), in relation to the isotope fraction, which reacts for each element within the system.

Within this model, under the same established conditions and considering an equilibrium constant KIE = −32, the results of the black mudstone (**Table 3**) also were

Δ1=	**39,**	Δ2 = **12,3**				
		Open system two products		**Older samples at top-inverse sequence**		
	H_2S	$CaCO_3$	CH_4	sample	$\delta^{13}C$ ‰(V-PDB)	Loss - gain
f	δ pool, H_2S	δ product 1, $CaCO_3$	δ product 2, CH_4	No	Lab results	
0	12,3	−26,7	0			13C enrichment 20‰
0,1	14,97	−24,03	2,67	11	−24,00	
0,2	17,64	−21,36	5,34			
0,3	20,31	−18,69	8,01	11a	**-18,00**	
0,4	22,98	−16,02	10,68	32c	**-13,31**	
0,5	25,65	−13,35	13,35	48	−8,70	
0,6	28,32	−10,68	16,02	45	**-5,36**	
0,7	30,99	−8,01	18,69	23	−12,78	13C depletion 7‰
0,8	33,66	−5,34	21,36	24	**-10,60**	
0,9	36,33	−2,67	24,03			
1	39	0	26,7			

Table 2.
Fractionation factors (bold) for CH_4 and $CaCO_3$ for green fossiliferous claystone.

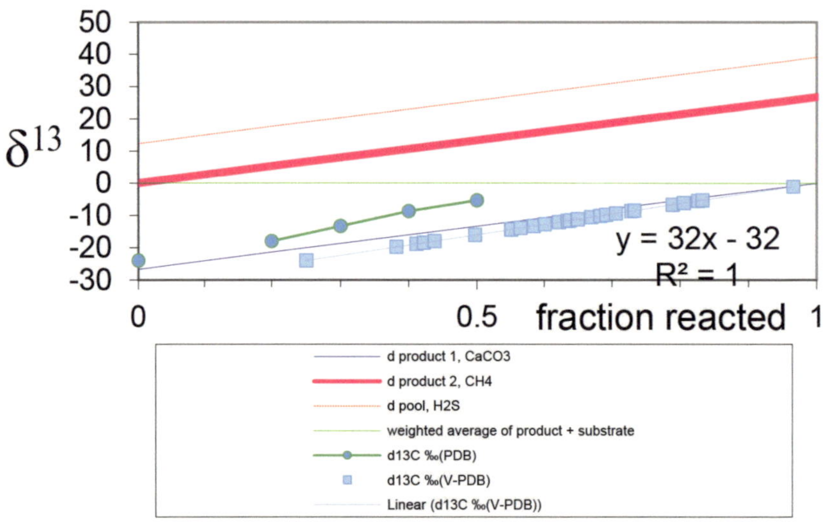

Figure 2.
Fractionation in an open system of Cerro Matoso hydrothermal sediments (green fossiliferous claystone), with exact equations for fractionation in the residual substrate (top line) and product (blue line $f = 1-(\delta^{13}C$ ‰$/-32)$. Green dotes when Δ1 = 39‰, y Δ2 = 12.3‰.

plotted (**Figure 3**). The $\delta^{13}C$ in V-PDB of the full 27 samples analyzed from Cerro Matoso (blue boxes) are represented in both figures. As a result, the $\delta^{13}C$ values from Cerro Matoso coincide with the ^{13}C isotope values expected for the precipitation of CaCO3 or product 1 (solid line) according to the model.

Δ1 =	**39**	**Δ2 = 12,3**				
		Open system two products		**Older samples at top-inverse sequence**		
	H2S	$CaCO_3$	CH_4	sample	$\delta^{13}C$ ‰(V-PDB)	Loss - gain
f	δ pool, H2S	δ product 1, $CaCO_3$	δ product 2, CH_4	No	Lab results	
0	12,3	−26,7	0	1	**-27,15**	13C enrichment 20‰
0,1	14,97	−24,03	2,67	35	−17,10	
0,2	17,64	−21,36	5,34	20	−21,63	
0,3	20,31	−18,69	8,01	26	−9,89	
0,4	22,98	−16,02	10,68	19	−9,39	
0,5	25,65	−13,35	13,35	52	−8,17	
0,6	28,32	−10,68	16,02	13	−10,54	13C depletion 2‰
0,7	30,99	−8,01	18,69	13	−12,11	
0,8	33,66	−5,34	21,36	46 ocre	**-10,09**	
0,9	36,33	−2,67	24,03	43	−11,55	
1	39	0	26,7			

Table 3.
Fractionation factors (bold) for CH4 and CaCO3 for black mudstone succession.

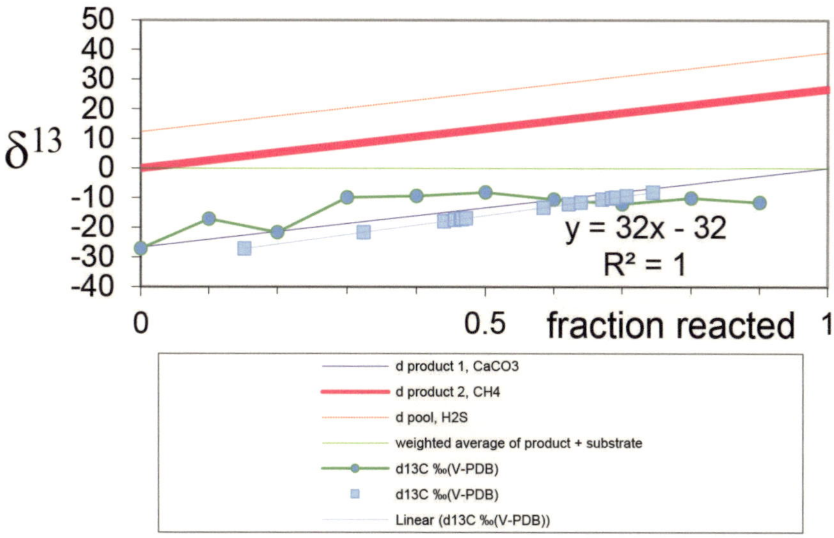

Figure 3.
Fractionation in an open system of Cerro Matoso metalliferous sediments (black mudstone succession), with exact equations for fractionation in the residual substrate (top line) and product (blue line $f = 1-(\delta^{13}C$ ‰$/-32)$. Green dots when $\Delta 1 = 39$‰, y $\Delta 2 = 12.3$‰.

The variations in time during the isotopic fractionation process can be analyzed in detail, on a medium or large scale. In the Cerro Matoso stratigraphic succession, this variation during mineral precipitation can be observed in the record of carbonates present in the samples. Considering the location of each sample in-depth, the formation time can be related to the hydrothermal activity. In principle, the data indicate that there was an approximate average enrichment in $\delta^{13}C$ of 20‰, in the first part of the succession, which is proposed as a stage for the formation of clay minerals. The samples of higher levels in the stratigraphic succession in the mounds, with averages $\delta^{13}C = -10‰$, suggest a 5 to 7‰ depletion of $\delta^{13}C$ in these rocks, a decrease that can occur in sediments located further from the vents or when a decrease in the intensity of these is represented (variations in contributions of CO_2 or CH_4).

The $\delta 1^{3}C$ values between (i.e., −24 to −16.9‰ in basal fissured and breccia mudstones - fb-M), suggest that the main carbon source is methane while the $\delta^{13}C$ ranges between (−14.4 to −10.1‰) and (−8.5 to −1.5‰) in the green fossiliferous claystone indicate episodes of dissolved inorganic carbon input— DIC that are mixed with methane sources. For their part, the carbonates that form the biostructures and concretions within the green fossiliferous claystone, with $\delta^{13}C$ values between (−9.4 to −6.2‰) and (−11.7 to −11.2‰), also suggest contributions of DIC and methane during their training.

Due to the fact that today the only carbonate present in the samples is siderite, it is possible to assume, according to the ^{13}C and ^{18}O isotope composition that siderite (diagenetic origin) replaced the primary fossils carbonate in the claystone, inheriting the isotopic composition of the precursor. This is considering the minimal difference (= 0.00042) in the fractionation factor between aragonite (Δ = 1.00856) and siderite (Δ = 1.00898) at 75°C [115], which indicates that the temperatures recorded by the $\delta^{18}O$, are quite consistent with the estimates of the initial fluid–rock interaction, in the hydrothermal system suggested for Cerro Matoso. Everything suggests that the siderite is formed in the process of early diagenesis in which, due to fluid–rock interaction on the seabed, Fe replaces Ca in the carbonate structure, so that the CO_3 ion preserves the isotopic footprint of the hydrothermal environment where the benthic community settled.

Accordingly, if the original authigenic carbonates inherited the $\delta^{13}C$ from the characteristic values of their specific carbon sources (+/− with DIC mixture), their $\delta^{13}C$ signature indicates the existence of microbial processes, which are characteristic in vent-related carbonate deposits [116]. Thus, the isotopic signature was preserved during the diagenetic processes and later during the supergene processes suffered by the Cerro Matoso sediments, the former, favored by the low temperatures of the ¨diffuse¨ fluids and the minimal variation in the fractionation factor between aragonite and siderite.

Owing to the isotopic fingerprinting of minerals formed under submarine conditions and the study of current hydrothermal systems, both the mid-oceanic ridge of the Atlantic and the mid-oceanic Pacific ridge, was possible to reconstruct the geological history of Cerro Matoso, and other deposits in ophiolites worldwide (e.g., sulfide ore of the Samal Ophiolite, Troodos ophiolite of Cyprus), suggesting that serpentinization-derived fluids flow are active for a considerable time during the formation of the mounds and tabular units, as well as, that the ocean is intimately related to the continents and that methane produced in active ancient ocean ridges appear to be similar to the methane produced in the actual mid-ocean ridge configuration.

Finally, many vent communities found on continents today must have their ancestors in ancient vent communities. In this sense, Kiel S [117] suggests that the fauna found today may, due to its exceptional level of endemism in a high taxonomic range, suggest a long and continuous evolutionary history that would extend from the Paleozoic in some parts of the world, where they have been reported.

DOI: http://dx.doi.org/10.5772/intechopen.110697

Author details

Andrés Castrillón
Universidad Nacional de Colombia, Bogotá

*Address all correspondence to: acastrillonp@unal.edu.co

References

[1] Yu BS, Chen JQ, Li XW, et al. The geochemistry of Lower Cambrian black shales and its relation to crustal evolution in the Tarim Basin. Science China (Ser D). 2002;**32**:374-382 [in Chinese]

[2] Yang R, Wei H, Bao M, Wang W, Wang Q, Zhang X, et al. Discovery of hydrothermal venting community at the base of Cambrian barite in Guizhou Province, Western China: Implication for the Cambrian biological explosion. Progress in Natural Science. 2008;**18**:65-70. DOI: 10.1016/j.pnsc.2007.07.006

[3] Kelley D, Shank T. Hydrothermal systems: A decade of discovery. In: Slow Spreading Environments in Diversity of Hydrothermal Systems on Slow Spreading Ocean Ridges. Geophysical Monograph Series. Washington, USA. Vol. 188. 2010. pp. 369-407

[4] Früh-Green GL, Kelley DS, Bernasconi SM, Karson JA, Ludwig KA, Butterfield DA, et al. 30,000 years of hydrothermal activity at the lost City vent field. Science Reports. 2003;**301**:495-498. Available from:http://science.sciencemag.org/

[5] Früh-Green GL, Connelly JA, Plas A, Kelley DS, Grobéty B. Serpentinization of oceanic peridotites: Implications for geochemical cycles and biological activity. In: The Subseafloor Biosphere at Mid-Ocean Ridges. Washington, D.C: American Geophysical Union; 2004

[6] Konn C. Origin of Organic Compounds in Fluids from Ultramafic Hosted Hydrothermal Vents of the Mid-Atlantic Ridge. Stockholm: Meddelanden från Stockholms Universitets Intitution för Geologi och Geokemi. 2009

[7] Hoefs J. Stable Isotope Geochemistry. Berlin: Springer-Verlag; 2015. p. 389

[8] Van Dover CL. Biogeography of hydrothermal vent communities along seafloor spreading centers. Trends in Ecology & Evolution. 1990;**5**:242-246

[9] Gebruk AV, Chevaldonne P, Shank T, Lutz RA, Vrijenhoek RC. Deep-sea hydrothermal vent communities of the Logatchev area (14°45'N, mid-Atlantic ridge): Diverse biotopes and high biomass. Journal of the Marine Biological Association of the UK. 2000;**80**:383-393

[10] Desbruyères D, Almeida A, Biscoito M, Comtet T, Khripounoff A, Le Bris N, et al. A review of the distribution of hydrothermal vent communities along the northern Mid-Atlantic Ridge: Dispersal vs. environmental controls. Hydrobiology. 2000;**440**:201-216

[11] Hessler RR, Smithey WM. The distribution and community structure of megafauna at the Galapagos Rift hydrothermal vents. In: Rona PA et al., editors. Hydrothermal Processes at Seafloor Spreading Centers. New York: Plenum; 1983. pp. 735-770

[12] Jorgensen BB, Boetius A. Feast and famine — Microbial life in the deep-sea bed. Focus on Marine Microbiology. 2007;**5**:770-781

[13] Delaney JR, Kelley DS, Lilley MD, Butterfield DA, Baross JA, Wilcock WSD, et al. The quantum event of oceanic crustal accretion: Impact of diking at Mid-Oceanic Ridges. Chemistry and Biology of the Oceans Science. 1998;**281**:222-230

[14] Heikinian R, editor. Petrology of the Ocean Floor. Elsevier Oceanography Series. New York: Elsevier scientific publishing company; 1982. p. 393

[15] Jannasch HW. The chemosynthetic support of life and the microbial

diversity at deep-sea hydrothermal vents. Proceeding of the Royal Society of London Biological Science. 1985;**225**:277-297

[16] Travis J. Probing de unsolved mysteries of the deep. Science. 1993;**259**:1123-1124

[17] Farmer JD. Hydrothermal systems: Doorways of the early biosphere evolution. GSA Today. 2000;**10**:1-9

[18] Davies P. The origin of the life I: When and where did it begin? Scientific Progress. 2001;**84**:1-16

[19] Campbell KA. Hydrocarbon seep and hydrothermal vent paleoenvironments and paleontology: Past developments and future research directions. Palaeogeography Palaeoclimatology Palaeoecology. 2006;**232**(2-4):362-407

[20] Haymon RM, Koski RA, Sinclair C. Fossils of hydrothermal vent worms from Cretaceous sulfide ores of the Samail ophiolite, Oman. Science. 1984;**223**:1407-1409

[21] Little CTS, Herrington RJ, Maslennikov VV, Morris NJ, Zaykov VV. Silurian hydrothermal-vent community from the southern Urals, Russia. Nature. 1997;**385**:146-148

[22] Little CTS, Can JR, Herrington RJ, Morisseau M. Late Cretaceous hydrothermal vent communities from the Troodos ophiolite, Cyprus. Geology. 1999;**27**:1027-1030

[23] Hovland JAG. Seabed Pockmarks and Seepages: Impact on Geology, Biology and the Marine Environment. London: Graham and Trotman; 1988. pp. 1-293

[24] Judd AG. The global importance and context of methane escape from the seabed. Geo-Marine Letters. 2003;**23**:147-154

[25] Kelley DS, Karson JA, Blackman DK, Früh-Green GL, Butterfield DA, Lilley MD, et al. AT3-60 shipboard party, an off-axis hydrothermal vent field near the Mid- Atlantic ridge at 308N. Nature. 2001;**412**:145-148

[26] Han X, Suess E, Sahling H, Wallman K. Fluid venting activity on the Costa Rica margin: New results from authigenic carbonates. International Journal of Earth Sciences. 2004;**93**:596-611

[27] Peckmann J, Little CTS, Gill F, Reitner J. Worm tube fossils from the Hollard Mound hydrocarbon-seep deposit, Middle Devonian, Morocco: Palaeozoic seep-related vestimentiferans? Palaeogeography Palaeoclimatology Palaeoecology. 2005;**227**:242-257

[28] Williams AB, Rona PA. Two new caridean shrimps (Bresilidae) from hydrothermal field on the Mid-Atlantic Ridge. Journal of Crustacean Biology. 1986;**6**:446-462

[29] Lutz RA, Kennish MJ. Ecology of deep-sea hydrothermal vent communities: A review. Reviews of Geophysics. 1993;**31**:211-242

[30] Shank TM, Fornarik DJ, Von Damm KL, Lilley MD, Haymon RM, Lutz RA. Temporal and spatial patterns of biological community development at nascent deep-sea hydrothermal vents (98N, East Pacific Rise). Deep Sea Research. 1998;**45**:465-516

[31] Tunnicliffe V, Juniper SK. Dynamic character of the hydrothermal vent habitat and the nature of the sulfide chimney fauna. Progress in Oceanography. 1990;**24**:1-14

[32] Castrillón A. Carbonatos y otros minerales autigénicos asociados a las lateritas niquelíferas de Cerro Matoso y su posible relación con actividad hidrotermal y reducción de sulfatos [Ph.D thesis]. Bogotá: Universidad Nacional de

Colombia; 2019. Available from: https://repositorio.unal.edu.co/handle/unal/79059

[33] Banerjee NR, Simonetti A, Furnes H, Muehlenbachs K, Staudigel H, Heaman L, et al. Geology. 2007;**35**:487

[34] Fisher C, Roberts H, Cordes E, Bernard B. Cold seeps and associated communities of the Gulf of Mexico. Oceanography. 2007;**20**(4)

[35] Niemann H, Lösekann T, de Beer D, Elvert M, Nadalig T, Knittel K, et al. Novel microbial communities of the Haakon Mosby mud volcano and their role as methane sink. Nature. 2006;**443**(7113):854-858

[36] Kulm LD, Suess E, Moore JC, Carson B, Lewis BT, Ritger SD, et al. Oregon subduction zone: Venting, fauna and carbonates. Science. 1986;**231**:561-566

[37] Bohrmann G, Greinert J, Suess E, Torres M. Authigenic carbonates from the Cascadia subduction zone and their relation to gas hydrate stability. Geology. 1998;**26**(7):647-650

[38] Boetius A, Ravenschlag K, Schubert CJ, Rickert D, Widdel F, Gieseke A, et al. A marine microbial consortium apparently mediating anaerobic oxidation of methane. Nature. 2000;**407**:623-626

[39] Peckmann J, Gischler E, Oschmann W, Reitner J. An Early Carboniferous seep community and hydrocarbon-derived carbonates from the Harz Mountains, Germany. Geology. 2001;**29**:271-274

[40] Suess E. Gashydrat—Eine Verbindung aus Methan und Wasser. Nova Acta Leopoldina NF. 2002a;**85**(323):123-146

[41] Suess E. The evolution of an idea: From avoiding gas hydrates to actively drilling for them. In: Achievements and opportunities of scientific ocean drilling. The legacy of the Ocean Drilling Program. JOIDES Journal. 2002b;**28**(1):45-50

[42] Núñez-Useche F, Canet C, Liebetraud V, Pi Puig T, Poncianoe AC, Alfonso P, et al. Redox conditions and authigenic mineralization related to cold seeps in Central Guaymas Basin, Gulf of California. Marine and Petroleum Geology. 2018;**95**:1-15. DOI: 10.1016/j.marpetgeo.2018.04.010

[43] Campbell KA, Francis DA, Collins M, Gregory MR, Nelson CS, Greinert J, et al. Hydrocarbon seep-carbonates of a Miocene forearc (East Coast Basin), North Island, New Zealand. Sedimentary Geology. 2008;**204**:83-105

[44] Van Dover CL. The Ecology of Deep-Sea Hydrothermal Vents. New Jersey: Princeton University; 2000. p. 424

[45] Suttner TJ, Hint O. Devonian scolecodonts from the Tyrnaueralm, Graz Palaeozoic, Austria. Memoirs of the Association of Australasian Palaeontologists. 2010;**39**:139-145

[46] Dubé TE. Tectonic significance of Upper Devonian igneous rocks and bedded barite, Roberts Mountains allochthon, Nevada, U.S.A. In: McMillan NJ, Embry AF, Glass DJ, editors. Devonian of the World (Volume II, Sedimentation), Canadian Society of Petroleum Geologists Memoir; 1988;**14**:235-249

[47] Von Bitter P, Scott SD, Schenk PE. Early Carboniferous low-temperature hydrothermal vent communities from New-foundland. Nature. 1990;**344**:145-148

[48] Gómez-Pérez I. An Early Jurassic deep-water stromatolitic bioherm related to possible methane seepage (Los

Molles Formation Neuque´n, Argentina). Palaeogeography, Palaeoclimatology, Palaeoecology. 2003;**201**:21-49

[49] Campbell KA, Farmer JD, Des Marais D. Ancient hydrocarbon seeps from the Mesozoic convergent margin of California: carbonate geochemistry, fluids and palaeoenvironments. Geofluids. 2002;**2**:63-94

[50] Hikida Y, Suzuki S, Togo Y, Ijiri A. An exceptionally well-preserved seep community from the Cretaceous Yezo fore-arc basin in Hokkaido, northern Japan. Palaeontological Research. 2003;**7**:329-432

[51] Schwartz H, Sample J, Weberling KD, Minisini D, Moore JC. An ancient linked fluid migration system: cold-seep deposits and sandstone intrusions in the Panoche Hills, California, USA. Geo-Marine Letters. 2003;**23**:340-350

[52] Goedert JL, Squires RL. Eocene deep-sea communities in localized limestones formed by subduction-related methane seeps, southwestern Washington. Geology. 1990;**18**:1182-1185

[53] Goedert JL, Thiel V, Schmale O, Rau WW, Michaelis W, Peckmann J. The Late Eocene ¨Whiskey Creek methane-seep deposit (western Washington State)—Part I: geology, palaeontology, and molecular geobiology. Facies. 2003;**48**:223-240

[54] Goedert JL, Campbell KA. An early Oligocene chemosynthetic community from the Makah Formation, northwestern Olympic Peninsula, Washington. Veliger. 1995;**38**:22-29

[55] Goedert JL, Peckmann J, Reitner J. Worm tubes in an allochthonous cold-seep carbonate from Lower Oligocene rocks of western Washington. Journal of Paleontology. 2000;**74**:992-999

[56] Naganuma T, Okayama Y, Hattori M, Kanie Y. Fossil worm tubes from the presumed cold-seep carbonates of the Miocene Hayama Group, central Miura Peninsula, Japan. The Island Arc. 1995;**4**:199-208

[57] Peckmann J, Thiel V, Michaelis W, Clari P, Gaillard C, Martire L, Reitner J. Cold seep deposits of Beauvoisin (Oxfordian; southeastern France) and Marmorito (Miocene; northern Italy): microbially induced authigenic carbonates. International Journal of Earth Sciences. 1999a;**88**:60-75

[58] Reysenbach AL, Cady SL. Microbiology of ancient and modern hydrothermal systems. Trends in Microbiology. 2001;**9**(2)

[59] Suess E, Carson B, Ritger S, Moore JC, Jones M, Kulm LD, et al., editors. Biological communities at vent sites along the subduction zones off Oregon. The hydrothermal vents of the Eastern Pacific: An overview. Bulletin Biological Society of Washington. 1985;**6**:475-484

[60] MacDonald IR, Bohrmann G, Escobar E, Abegg F, Blanchon P, Blinova V, et al. Asphalt volcanism and chemosynthetic life in the Campeche Knolls, Gulf of Mexico. Science. 2004;**304**:999-1002

[61] Lonsdale P. A deep-sea hydrothermal site on a strike-slip fault. Nature. 1979;**281**:531-534

[62] Lavoie D, Chi G. An Ordovician "Lost City"—Venting serpentinite and life oases on Iapetus seafloor. Canadian Journal of Earth Sciences. 2010;**47**(3):199-207. DOI: 10.1139/E10-013

[63] Schwarzenbach EM, Früh-Green GL, Bernasconi SM, Alt JC, Shanks III WC, Gaggero L, et al. Sulfur geochemistry of

peridotite-hosted hydrothermal systems: Comparing the Ligurian ophiolites with oceanic serpentinites. Geochimica et Cosmochimica Acta. 2012;**91**:283-305. DOI: 10.1016/j.gca.2012.05.021

[64] Hodel F, Macouin M, Trindade RIF, Triantafyllou A, Ganne J, Chavagnac V, et al. Fossil black smoker yields oxygen isotopic composition of Neoproterozoic seawater. Nature Communications. 2018;**9**:1453. DOI: 10.1038/s41467-018-03890-w

[65] Karson JA, Dick HJB. Tectonics of ridge-transform intersections at the Kane fracture zone. Marine Geophysical Researches. 1983;**6**(1):51-98

[66] Dick HJ, Meyer PS, Bloomer S, Kirby S, Stakes D, Mawer C. Lithostratigraphic evolution of an in-situ sections of ocean layer 3. In: Proceeding of the Ocean Drilling Program, Scientific Results. College Station, Texas. 1991;**118**:439-538

[67] Cannat M, Sauter D, Mendel V, Ruellan E, Okino K, Escartin J, et al. Modes of seafloor generation at a melt-poor ultraslow-spreading ridge. Geology. 2006;**34**:605-608. DOI: 10.1130/G22486.1

[68] Karson JA, Früh-Green GL, Kelley DS, Williams EA, Yoerger DR, Jakuba M. Detachment shear zone of the Atlantis Massif core complex, Mid-Atlantic Ridge, 30 N. Geochemistry, Geophysics, Geosystems. 2006;**7**(6):1-29. DOI: 10.1029/2005GC001109

[69] Cann JR, Blackman DK, Smith DK, McAllister E, Janssen B, Mello S, et al. Corrugated slip surfaces formed at ridge-transform intersections on the Mid-Atlantic Ridge. Nature. 1997;**385**:329-332. DOI: 10.1038/385329a0

[70] Escartín J, Cannat M. Ultramafic exposures and the gravity signature of the lithosphere near the Fifteen Twenty Fracture Zone (Mid Atlantic Ridge, 14-16.5N). Earth Planter Science Letters. 1999;**171**:411-424

[71] Smith DK, Cann JR, Escartin J. Widespread active detachment faulting and core complex formation near 13° N on the Mid-Atlantic Ridge. United Kingdom. Nature. 2006;**442**:440-443. DOI: 10.1038/nature04950

[72] Silantyev SA, Mironenko MV, Novoselov AA. Hydrothermal systems in peridotites of slow-spreading mid-oceanic ridges. Modeling phase transitions and material balance: Down-welling limb of a hydrothermal circulation cell. Petrology. 2009;**17**(2):138-157

[73] Kelley DS, Baross JA, Delaney JR. Volcanoes, fluids, and life at mid-ocean ridge spreading centers. Annual Review of Earth and Planetary Science. 2002;**30**(1):385-491

[74] German CR, Von Damm KL. Hydrothermal processes. In: Elderfield H, editor. The Oceans and Marine Geochemistry: Treatise on Geochemistry. Vol. 6. Elsevier Ltd.; 2003;**6**:181-222

[75] Alt J, Shanks W. Serpentinization of abyssal peridotites from the MARK area, Mid Atlantic Ridge: Sulfur geochemistry and reaction modeling. Geochimica et Cosmochimica Acta. 2003;**67**:641-653

[76] Alt JC, Shanks III WC, Bach W, Paulick H, Garrido CJ, Beaudoin G. Hydrothermal alteration and microbial sulfate reduction in peridotite and gabbro exposed by detachment faulting at the Mid Atlantic Ridge, 15o20′N (ODP Leg 209): A sulfur and oxygen isotope study. Geochemistry, Geophysics, Geosystems. 2007;**8**(8). DOI: 10.1029/2007GC001617

[77] Delacour A, Früh-Green G, Bernasconi MS. Sulfur mineralogy and

geochemistry of serpentinites and gabbro of the Atlantis Massif (IODP Site U1309). Geochimica et Cosmochimica Acta. 2008;**72**:5111-5127

[78] Ildefonse B, Rona PA, Blackman D. Drilling the crust at Mid-Ocean Ridges. An in-depth perspective. Oceanography. 2007;**20**(1):66-77

[79] Sinton CW, Duncan RA, Storey M, Lewis J, Estrada JJ. An oceanic flood basalt province within the Caribbean Plate. Earth and Planetary Science Letters. 1998;**155**:221-235

[80] Pindell J, Barrett S. Geological evolution of the Caribbean region: A plate tectonic perspective. In: Dengo G, Case JE, editors. The Caribbean Region. Boulder, Colorado, USA: Geological Society of America. 1991. DOI: 10.1130/DNAG-GNA-H

[81] Pindell JL, Kennan L. Tectonic Evolution of the Gulf of Mexico, Caribbean and Northern South America in the Mantle Reference Frame: An Update. Vol. 328. London: Geological Society; 2009. pp. 1-55. DOI: 10.1144/SP328.1

[82] Lewis J, Draper G, Proenza J, Epaillat J, Jiménez J. Ophiolite-related ultramafic rocks (Serpentinites) in the Caribbean region: A review of their occurrence, composition, origin, emplacement and Ni-laterite soil formation. Geologica Acta. 2006;**4**(1-2): 237-263. DOI: 10.1344/105.000000368

[83] Nívia A. El Complejo Estructural Dagua, Registro de Deformación de la Provincia Litosférica Cretácica Occidental en un Prisma Acrecionario. VII Congreso Colombiano de Geología Tomo III. 1997. pp. 54-67

[84] Wierzbowski H, Joachimski M. Reconstruction of late Bajocian-Bathonian marine paleoenvironments using carbon and oxygen isotope ratios of calcareous fossils from the Polish Jura Chain (Central Poland). Palaeogeography, Palaeoclimatology, Palaeoecology. 2007;**254**:523-540. DOI: 10.1016/j.palaeo.2007.07.010

[85] Macgregor D. Relationships between seepage, tectonics and subsurface petroleum reserves: Marine and Petroleum Geology. 1993;**10**:606-619

[86] Abrajano TA, Sturchio NC, Bohlke JK, Lyon GL, Poreda RJ, Stevens CM. Methane-hydrogen gas seeps, Zambales ophiolite, Philippines: Deep or shallow origin? Chemical Geology. 1988;**71**(3):211-222. DOI: 10.1016/0009-2541(88)90116-7

[87] Lilley MD, Butterfield DA, Olson EJ, Lupton JE, Macko SA, McDuff RE. Anomalous CH4 and NH4+ concentrations at an unsedimented mid-ocean-ridge hydrothermal system. Nature. 1993;**364**(6432):45-47. DOI: 10.1038/364045a0

[88] Charlou JL, Donval JP, Fouquet Y, Jean-Baptiste P, Holm N. Geochemistry of high H2 and CH4 vent fluids issuing from ultramafic rocks at the rainbow hydrothermal field (36°14′N, MAR). Chemical Geology. 2002;**191**(4):345-359. DOI: 10.1016/S0009-2541(02)00134-1

[89] Proskurowski G, Lilley MD, Olson EJ. Stable isotopic evidence in support of active microbial methane cycling in low-temperature diffuse flows vents at 9°50′N East Pacific rise. Geochimica et Cosmochimica Acta. 2008a;**72**(8):2005-2023. DOI: 10.1016/j.gca.2008.01.025

[90] Meister P, Wiedling J, Lott C, Bach W, Kuhfuß H, Wegener G, et al. Anaerobic methane oxidation inducing

carbonate precipitation at abiogenic methane seeps in the Tuscan archipelago (Italy). PLoS One. 2018;**13**(12):1-34. DOI: 10.1371/journal.pone.0207305

[91] Sciarra A, Saroni A, Etiope G, Coltorti M, Mazzarini F, Lott C, et al. Shallow submarine seep of abiotic methane from serpentinized peridotite off the Island of Elba, Italy. Applied Geochemistry. 2019;**100**:1-7. DOI: 10.1016/j.apgeochem.2018.10.025

[92] Eickmann B, Little CTS, Peckmann J, Taylor PD, Boyce AJ, Morgan DJ, et al. Shallow-marine serpentinization-derived fluid seepage in the Upper Cretaceous Qahlah Formation. United Arab Emirates. Geological Magazine. 2021;**158**(9):1561-1571. DOI: 10.1017/S0016756821000121

[93] Douville E, Charlou JL, Oelkers EH, Bienvenu P, Colon CJ, Donval JP, et al. The rainbow vent fluids (36 14′ N, MAR): The influence of ultramafic rocks and phase separation on trace metal content in Mid-Atlantic Ridge hydrothermal fluids. Chemical Geology. 2002;**184**(1-2):37-48. DOI: 10.1016/S0009-2541(01)00351-5

[94] Kelley DS, Karson JA, Früh-Green GL, Yoerger DR, Shank TM, Butterfield DA, et al. A serpentinite-hosted ecosystem: The Lost City hydrothermal field. Science. 2005;**307**(5714):1428-1434

[95] Fu Q, Lollar BS, Horita J, Lacrampe-Couloume G, Seyfried WE Jr. Abiotic formation of hydrocarbons under hydrothermal conditions: Constraints from chemical and isotope data. Geochimica et Cosmochimica Acta. 2007;**71**(8):1982-1998

[96] Lartaud F, de Rafélis M, Oliver G, Krylova E, Dyment J, Ildefonse B, et al. Fossil clams from a serpentinite-hosted sedimented vent field near the active smoker complex rainbow, MAR, 36o13'N: Insight into the biogeography of vent fauna. Geochemistry, Geophysics, Geosystems. 2010;**11**(8):1-17. DOI: 10.1029/2010GC003079

[97] Castrillón A, Pi-Puig T, Guerrero J, Nuñez-Useche F, Rodriguez A, Canet C. Clay mineralogy and texture of deep-sea hydrothermal mudstone associated with the Cerro Matoso peridotite in accreted oceanic crust from Colombia. Journal of South American Earth Sciences. 2022;**117**:1-17. DOI: 10.1016/j.jsames.2022.103886

[98] Klein F, Humphris SE, Guo W, Schubotz F, Schwarzenbach EM, Orsi WD. Fluid mixing and the deep biosphere of a fossil Lost City-type hydrothermal system at the Iberia Margin. Proceedings of the National Academy of Sciences of the United States of America. 2015;**112**(39):12036-12041. DOI: 10.1073/pnas.1504674112

[99] Carothers WW, Adami LH, Rosenbauer RJ. Experimental oxygen isotope fractionation between siderite-water and phosphoric acid liberated CO2-siderite. Geochimica et Cosmochimica Acta. 1988;**52**(10):2445-2450. DOI: 10.1016/0016-7037(88)90302-X

[100] Savin SM, Epstein S. The oxygen and hydrogen isotope geochemistry of clay minerals. Geochimica et Cosmochimica Acta. 1970;**34**(1):25-42. DOI: 10.1016/0016-7037(70)90149-3

[101] Gaudin A, Decarreau A, Noack Y, Grauby O. Clay mineralogy of the nickel laterite ore developed from serpentinised peridotites at Murrin Murrin, Western Australia. Australian Journal of Earth Sciences. 2005;**52**(2):231-241

[102] Irwin H, Curtis C, Coleman M. Isotopic evidence for source of diagenetic

carbonate forms durign burial of organic-rich sediments. Nature. 1977;**269**:209-213

[103] Palandri JL, Reed MH. Geochemical models of metasomatism in ultramafic systems: Serpentinization, rodingitization, and sea floor carbonate chimney precipitation. Geochimica et Cosmochimica Acta. 2004;**68**(5):1115-1133. DOI: 10.1016/j.gca.2003.08.006

[104] Proskurowski G, Lilley M, Seewald JS, Früh-Green GL, Olson EJ, Lupton JE, et al. Abiogenic hydrocarbon production at Lost City hydrothermal field. Science. 2008b;**319**(5863):604-607. DOI: 10.1126/science.1151194

[105] Lartaud F, De Rafélis M, Little CTS, Dyment J, Bayon G, Ildefonse B, et al. Ultramafic hydrothermal systems on the rainbow abyssal hill: A wide variety of active and fossil chemosynthetic habitats. InterRidge News. 2011;**20**:32-36

[106] Dias Á, Barriga F. Mineralogy and geochemistry of hydrothermal sediments from the serpentinite-hosted Saldanha hydrothermal field (36°34'N, 33°26'W) at MAR. Marine Geology. 2006;**225**(1-4):157-175. DOI: 10.1016/j.margeo.2005.07.013

[107] Charlou J, Fouquet Y, Bougault H, Donval J, Etoubleau J, Jean-Baptiste P, et al. Intense CH_4 plumes generated by serpentinization of ultramafic rocks at the intersection of the 15o20′N fracture zone and the Mid-Atlantic Ridge. Geochimica et Cosmochimica Acta. 1998;**62**:2323-2333

[108] Charlou JL, Donval JP, Konn C, Ondreas H, Fouquet Y, Jean-Baptiste P, et al. High production and fluxes of H2 and CH4 and evidence of abiotic hydrocarbon synthesis by serpentinization in ultramafic-hosted hydrothermal systems on the Mid-Atlantic Ridge. In: Rona P, et al., editors. Diversity of Hydrothermal Systems on Slow-Spreading Ocean Ridges. Washington, D. C: AGU; 2010;**188**

[109] Campbell AR, Larson PB. Introduction to stable isotope applications in hydrothermal systems. In: Richards JP, Larson AR, editors. Techniques in Hydrothermal Ore Deposits Geology. Littleton: Society of Economic Geologists; 1998

[110] Alt J, Muehlenbachs K, Honnorez J. An oxygen isotopic profile through the upper kilometer of the oceanic crust, DSDP Hole 504B. Earth and Planetary Science Letters. 1986;**80**(3-4):217-229. DOI: 10.1016/0012-821X(86)90106-8

[111] Hoefs J, Sywall M. Lithium isotope composition of quaternary and tertiary biogene carbonates and a global lithium isotope balance. Geochimica et Cosmochimica Acta. 1997;**61**(13):2679-2690. DOI: 10.1016/S0016-7037(97)00101-4

[112] Whiticar MJ, Faber E. Methane oxidation in sediment and water column environments—Isotopic evidence. Organic Geochemistry. 1986;**10**(4-6):759-768. DOI: 10.1016/S0146-6380(86)80013-4

[113] Fry B. Stable Isotope Ecology. New York: Springer; 2008

[114] Templeton AS, Staudigel H, Tebo BM. Diverse Mn (II)-oxidizing bacteria isolated from submarine basalts at Loihi Seamount. Geomicrobiology Journal. 2005;**22**(3-4):127-139. DOI: 10.1080/01490450590945951

[115] Kim S-T, Mucci A, Taylor BE. Phosphoric acid fractionation factors for calcite and aragonite between 25 and 75 °C. Chemical Geology. 2007;**246**:135-146

[116] Aharon P, Graber ER, Roberts HH. Dissolved carbon and $\delta^{13}C$ anomalies in

the water column caused by hydrocarbon seep on the northwestern Gulf of Mexico slope. Geo-Marine Letters. 1992;**12**:33-40

[117] Kiel S, Dando PR. Chaetopterid tubes from vent and seep sites: Implications for fossil record and evolutionary history of vent and seep annelids. Acta Palaeontologica Polonica. 2009;**54**(3):443-448. DOI: 10.4202/app.2009.0022

Section 3

Interaction of Oceans with Continents

Chapter 4

Large-Scale Interaction of Oceans, Continents and the Atmosphere in the Northern Hemisphere

Alexander Roshdestvensky

Abstract

The main interaction between oceans and continents is due to the atmosphere. As a result of this interaction, winds, heat flows and masses are generated between continents and oceans. In this section, we will touch on a little-covered topic about the largest thermal interaction of oceans, continents, and the atmosphere in terms of energy, including those arising due to seasonal fluctuations in their temperatures caused by the annual course of solar radiation. To answer the question about the magnitude and structures of seasonal and climatic heat transfer over oceans and continents, we will consider two components of heat and moisture flows from the ocean to the atmosphere—one component arises solely due to seasonal fluctuations in parameters (temperatures and wind), the other due to conditionally constant wind temperatures throughout the years. Further, we will discover the structures of global heat transfers by studying the phase mismatch of the temperature and pressure fields of the atmosphere. A tool for conducting such a phase analysis will be a shift phase, which detects a shift between two similar functions on a segment. This research program requires many explanations, which we will consistently provide below.

Keywords: ocean, atmosphere, continent, seasonal oscillatory heat flow, natural heat pump, climate

1. Introduction

A tool for calculations for turbulent heat and moisture flows at the water–air interface is the semi-empirical relations (1):

$$Q = c_p \rho c_h U(Tw - Ta), Q_e = \rho_a L_e c_e \mathrm{U}(\mathrm{e}_0 - \mathrm{e}_z), \tag{1}$$

where c_h, c_e are the coefficients of turbulent heat and moisture exchange in the air, ρ and c_p are the density and heat capacity of the air, L_e is the specific heat of evaporation, e_0, e_z are the actual humidity of the atmosphere and the absolute humidity of air saturation at water temperature, U is module of the driving wind, Q_h, Q_e are heat and moisture flows. These ratios are commonly called bulk formulas. Heat flow maps according to the calculations of (1) are given in many publications and reviews.

Detailed maps of heat flows according to bulk formulas (1) are given in [1, 2]. Flow calculations differ by 15–40% from different authors. Today, radiation measurements of fluxes from space are considered the most promising, but they also give a spread of flux values of 20–30% (**Figure 1**).

Formulas (1) are derived under the condition that there are no horizontal temperature gradients of water and air, which are always present in the ocean and atmosphere. A different parametrization of large-scale heat fluxes taking into account horizontal gradients is given in [3] to calculate integral heat fluxes for an annual period, in which the ideas of the relations (1) were used. For this, general assumptions were made that all parameters in (1), temperature, humidity and wind (U) were conditionally represented as a sum of two components, such as water temperature, where, is the annual period. Further, the large-scale portable wind was expressed in terms of geostrophic wind, depending on atmospheric pressure gradients. The key point was the author's idea to use atmospheric pressure as the sum of two terms, where the pressure of dry air, is the partial pressure of water vapor (humidity). The use of these assumptions and Bowen's relation on the similarity of heat and moisture fluxes made it possible to obtain simple formulas for calculating the integral annual heat fluxes at the water-air boundary in the form of (2) [3]:

$$\begin{aligned} &\tilde{Q} = \alpha_T \oint_{t_0} T_w dT_a, \tilde{Q}_e = \alpha_e \oint_{t_0} e_0 de_z, \overline{Q}_T = A_T\left(const\nabla\overline{T}_a + \nabla e_z\right)\left(\overline{T}_w - \overline{T}_a\right), \\ &\overline{Q}_e = A_e\left(const\nabla\overline{T}_a + \nabla e_z\right)\left(\overline{e}_0 - \overline{e}_z\right), \end{aligned} \quad (2)$$

where

$\tilde{Q}_T, \tilde{Q}_e$ are unidirectional flows of explicit and latent heat on average per year due to fluctuations T_a, T_w, e_0, e_z, where $\overline{Q}_T, \overline{Q}_e$ are flows of heat and moisture due to constant climatic values throughout the year T_a, T_w, e_0, e_z, and multipliers $\alpha_q, \alpha_e, A_T, A_e$ are dimensional constants. The analytical conclusion of the integral formulas (2) can be obtained independently of the bulk formulas (1) on the basis of ideas about the weak thermodynamic memory of the water-air system or on the basis

Figure 1.
It follows from the relations (1) that extreme heat flows occur at a strong speed of the portable wind (western transfer in the atmosphere) blowing from the continent to the ocean, the greatest temperature differences are water-air in the transition zone ocean-continent.

of the connection of horizontal gradients of water and air with vertical gradients of atmospheric temperature over the ocean.

The calculated heat and moisture fluxes according to (2) showed a good correspondence with the flow calculations according to (1). In (2) we obtain an amazing property of large-scale thermal interaction between the ocean and the atmosphere. When the constant climatic temperatures of water and air are equal, the heat flow between the ocean and the atmosphere, formed due to the difference in constant temperatures, stops. But in this case, a different component of the heat flow may remain due to fluctuations in these temperatures near the average value.

This implies the assumption that the greatest seasonal-oscillatory heat flows are where the greatest seasonal temperature fluctuations are observed, i.e. in the mid-latitude areas of the ocean. Calculations confirmed this hypothesis [4]. **Figure 2** show the distribution of integrals $Q = cons\oint T_w dT_a$ (see formula 2) in the ocean as heat flows. The geometric interpretation of the integral is the polar area of the temperature hysteresis loop (S) on the graph of the dependence or in annual temperature fluctuations, and. Therefore, **Figure 2** gives the distribution of temperature loops, the dimension of which $S = \oint_{t_0} T_w dT_a$, that is, $\left(\text{grad}^0 C\right)^2$, see formula (2).

Figure 3 shows seasonally-fluctuating heat flows. The shaded areas show the reverse heat flows from the ocean to the atmosphere. **Figure 3** shows seasonally-fluctuating heat flows.

The distribution of seasonal climatic and climatic heat fluxes determined by formulas (2) in the same ocean area is different. **Figure 4** shows the total flows of apparent and latent heat in the North Atlantic.

The seasonal component of heat ha **Figure 4** is "tied" to the ocean-continent boundary zone and does not have such a gap along the Gulf Stream as the climatic component. Seasonal heat flow does not have a maximum in the tropical zone, where

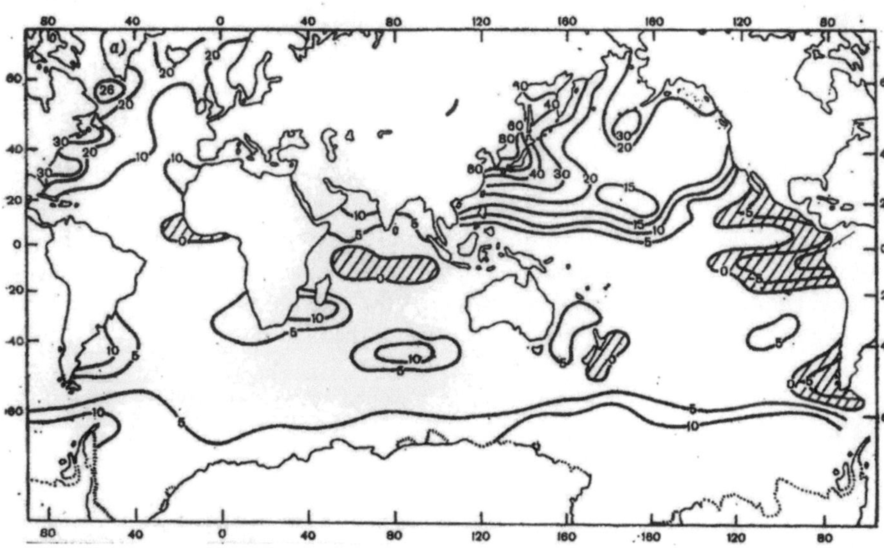

Figure 2.
Distribution in the world ocean the value of integrals $S = \oint_{t_0} T_w dT_a$ *(see formula 2) as the structure of the seasonal heat flow from the ocean to the atmosphere over an annual period.*

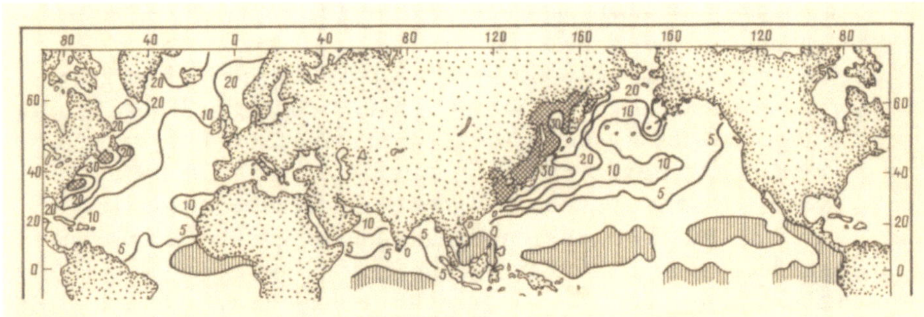

Figure 3.
Distribution of unidirectional "seasonal-oscillatory" heat fluxes $\tilde{Q} = \alpha_T \oint_{t_0} T_w dT_a$ *(see formula 2) per year. Areas of negative flows with values (−4 to 0) are shaded. In these areas, condensation of moisture prevails over the evaporation, of the double-hatching area and extremely high fluxes into the atmosphere.*

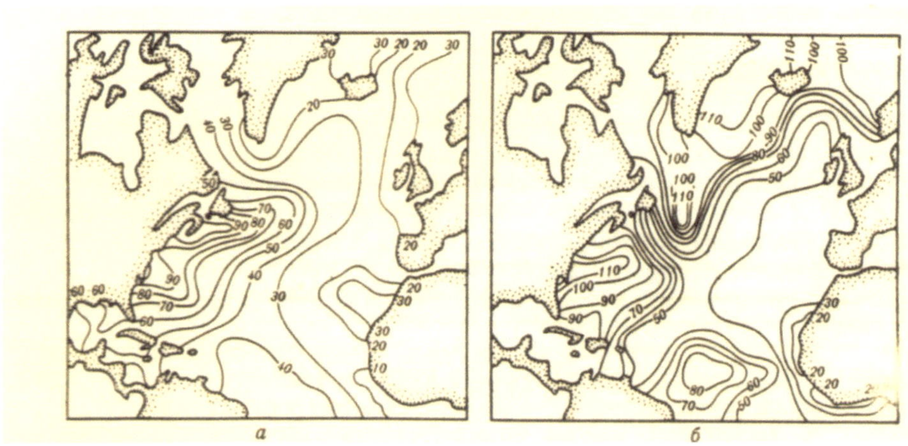

Figure 4.
The annual average seasonal (a) and climatic (b) components of the total heat transfer (explicit and latent heat) from the surface of the North Atlantic.

fluctuations in temperature and humidity throughout the year are minimal. The climatic heat flow is associated not only with the proximity of the continent but more with the Gulf Stream itself. It is assumed that the sharp break in the maximum of the climatic heat flow in Newfoundland is caused by the complex hydrology of the large-scale flow. Here, large volumes of cold waters flow into the Gulf Stream from the north, large-scale deepening of waters occurs and the beginning of the Gulf Stream branching.

The unidirectional seasonal-oscillatory transfer of apparent and latent heat into the Earth's atmosphere averaged over latitudinal zones per year is shown in **Figure 5** [5]. The maximum seasonal heat flows occur in the middle latitudes of the northern hemisphere, although the area of the oceans in the southern hemisphere is larger than in the northern. With a degree of simplification, it can be said that temperature fluctuations in temperatures and humidity on Earth, due to seasonal fluctuations in solar radiation and the distribution of continents, heat the northern hemisphere more than the southern hemisphere. Hence, the atmosphere of the northern hemisphere is on average 2.3 warmer than the southern hemisphere [6], and in the absence of

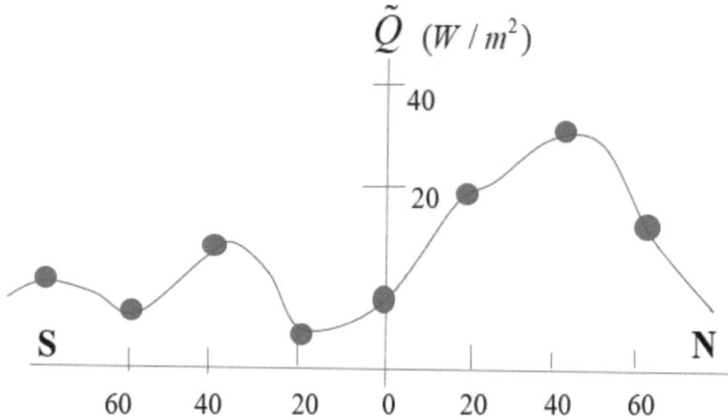

Figure 5.
The average annual zonal heat flow from the ocean to the atmosphere in the hemispheres of the Earth, including explicit and latent heat, W/m^2.

seasonal heat transfer, the average surface temperature of the continents in the northern hemisphere at a value of 5 would be 1.3 lower. There is no doubt that a seemingly insignificant increase in the temperature of the atmosphere and the surface of the continents in the northern hemisphere, due to seasonal heat flows, leads to an increase in the mass of the biosphere and the diversity of its species, especially in the middle latitudes of the northern hemisphere of the Earth.

2. The structure of divergence of heat fluxes in the atmosphere of the northern hemisphere as a structure of climatic zones of heating and cooling of the atmosphere

The main thermal energy comes from the ocean to the atmosphere from the surface of relatively localized zones—the so-called energy-active areas of the ocean. The question arises how and where this heat spreads in the atmosphere. To answer this question, the divergence of heat fluxes on the plane of each surface of the atmosphere, in the surface layer and on standard geopotential surfaces of the atmosphere at the level of 850, 700, 500, 300, 200, 100, and 50 mb. Heat transfer was taken into account by atmospheric geostrophic wind, which accounts for at least 90% of all portable air movements. These works [7], carried out in the northern hemisphere as a whole on climate data in the geostrophic approximation, for the first time showed that horizontal heat flows in the atmosphere have stable zones of divergence and convergence of heat flows in the ocean-continent transition zones, which are associated as sources of heat and "cold" in the atmosphere [7]. The divergence calculation was carried out using the Stokes theorem. To obtain the structure (map) of heat flows in the atmosphere, the heat flow carried out by air movements through the lateral boundaries in each isolated elementary flat block of the atmosphere with a size of 5x5 degrees of the computational grid was calculated. Then, to verify the results, a similar calculation was repeated on a smaller-scale grid, but the results of the calculations did not change much.

A map of the heat flux divergence isolines was constructed on each isobaric surface. Positive values of divergence correspond to the removal of heat from the circuit (heat source), and negative values correspond to the introduction of heat or its absorption zone. As a result, digital maps of heat fluxes for the northern hemisphere were obtained on 8 isobaric surfaces for each of the 12 climatic months, as well as on average by season and in general for the year, including integral data in the surface layer of the atmosphere (136 maps). These cards carry previously unknown information, but it is not possible to show them with a cavity in the chapter. We will cite those that prove the existence of only two large-scale structures in the system of horizontal heat transfer in the atmosphere of the northern hemisphere. We refused to use data with a higher resolution, calculations were made based on the ratios (3). The horizontal heat transfer (Q) through the side surface of one elementary plane contour of the atmosphere (L) of unit "thickness" on the geopotential surface is equal to:

$$Q = C_p h \oint_L \rho(x,t) T(x,t) \vec{v}(x,t) dL W/m^2 \tag{3}$$

where Cp is the heat capacity, h = 1 m is the thickness of the atmospheric layer 1 m, p is the air density, T is the temperature, is the horizontal air velocity on the horizontal contour L, and dL is the element of the contour. The normal component of the geostrophic air velocity to the contour, taking into account the Coriolis parameter at different latitudes, is equal to

$$(V_g)_n = \left({}^1\!/_{\rho f}\right)\left({}^{\partial P}\!/_{\partial L}\right) [m/\sec], \tag{4}$$

where $(V_g)_n$ is the normal component of the geostrophic wind velocity to the contour L, f is the Coriolis parameter P is atmospheric pressure.

To calculate the speed in the atmospheric layer 1015–850 mb. A constant multiplier Cs is added to expression (4)—a dimensionless multiplier reflecting the change in geostrophic wind in the surface and boundary layer up to heights of 700 m. The effect of this friction on the change in wind direction in the applied method of calculating the divergence of the flow can be ignored (see below). Consequently, the heat transfer through the contour (L) in the geostrophic approximation, taking into account (3) and (4), has the form (5):

$$Q = C_p \oint_L \rho T \vec{V}_n d\vec{L} = \left(\frac{C_p}{f}\right) \oint_L T \frac{\partial P}{\partial L} dL \; [W/m^2], \tag{5}$$

where Cp is the heat capacity of the air, Formula (5) was used to calculate heat fluxes at altitudes above 850 mb, so there is no correction factor for geostrophic wind.

If the heat transfer Q in (3) is represented in the coordinates of temperatures T and geopotential (Z) on an isobaric surface through a contour (L), then by analogy with the derivation of expressions (4) and (5), taking into account the definition $\frac{1}{\rho}\nabla P = \nabla Z$. we obtain an expression for heat transfer through the surface of the contour (L) in the coordinates of temperature (T) and geopotential (Z) through the contour L (6):

$$Q = Const \oint_L T \frac{\partial Z}{\partial L} dL \; [\mathrm{W/m^2}] \tag{6}$$

where Const = (Cp/fg), g is the acceleration of gravity, Cp is the heat capacity of the air, and formula (6) was used to calculate heat flows at altitudes above 850 mb, so there is no correction factor for geostrophic wind.

Consider the location of large-scale zones of heating and cooling of the atmosphere in winter and summer, **Figure 6**.

The results of calculations displayed on the map of the northern hemisphere revealed a rather unexpected, but quite understandable picture. Unexpected is the fact that the zones of heat removal and nose in the atmosphere, as volumetric sources and "heat sinks", form a structure on the map that does not fundamentally repeat the contours of continents and the ocean, and are located mainly in large-scale ocean–continent transition zones. It can be seen that the zones of divergence of large-scale heat fluxes in the atmosphere do not form a mosaic picture on the map but will unite into stable and homogeneous regions in sign [7], **Figure 6**. These regions are approximately similar to the geopotential disturbance zones in large-scale Wallace waves [8].

Figure 6 shows a map of heat transfer on the surface of 850 mb on average for the summer and winter periods and for the year. The average structure for the year is similar to the winter structure of heat transfer for a reason. In the winter structure, the regions of atmospheric heating have such a high power that they are clearly

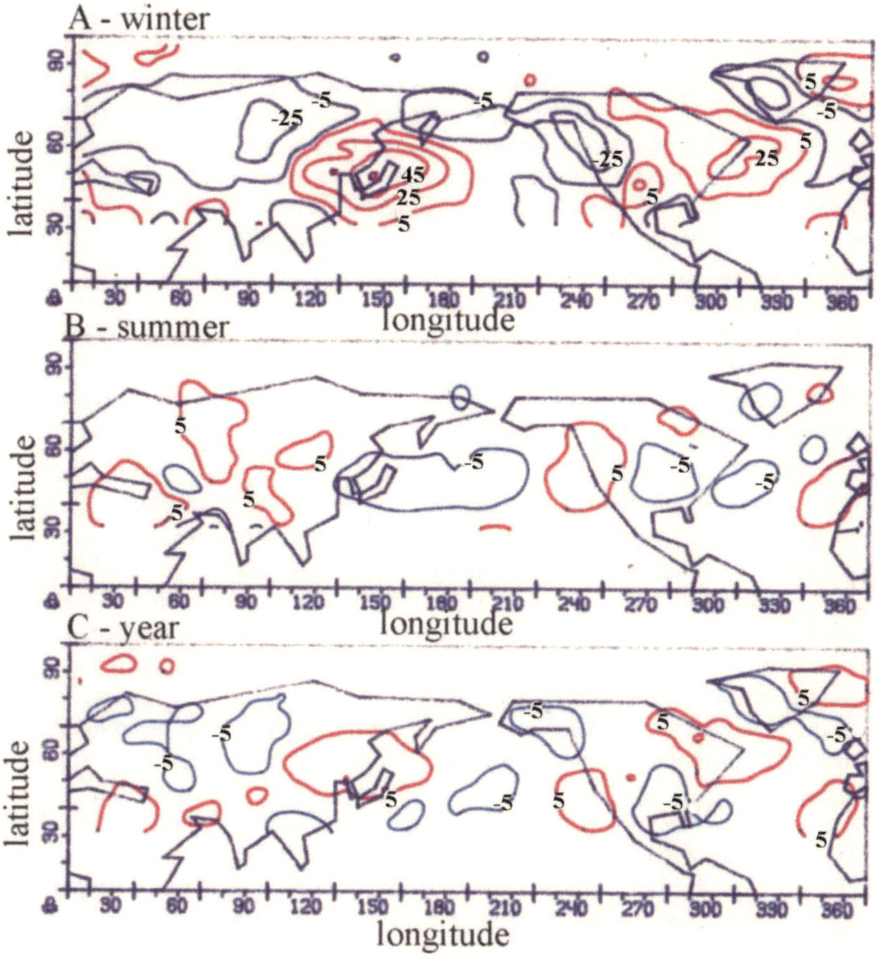

Figure 6.
Zones of divergence of heat flow in the atmosphere of the northern hemisphere as sources of heat and cold. The blue zone is radiators or sources of "cold". Red zones are heat sources. The surface of the atmosphere is 850 mb. Energy flow W/m².

manifested in the average annual structure. The summer structure is clearly visible here, in which the atmosphere heaters are the continent of Eurasia and the western regions of America and Africa with water areas, and the central regions of the oceans act as refrigerators.

Figure 7 also confirms the hypothesis that there are only two structures in the atmosphere of the heating and cooling zones of the atmosphere, winter and summer.

In **Figure** 7, the map represents the sources and sinks of heat in the atmosphere in an average year on the surface of 1013 MB, 850 MB, 700 MB, 500 MB, and 300 MB. The red counter marks the heat, and the black contour limits the zone of heat absorption. The most powerful heat area in the atmosphere is situated on the border of the continent of Asia and the ocean manifests at all altitudes.

Figure 8 shows maps of the divergence of the heat flow on an isobaric surface of 850 mb in the seasonal course for 6 months from summer to winter, from July to December. In July, the "cold" zones in the centre of the Pacific Ocean are clearly visible and the cooling zone (heat absorption) in the Atlantic is less intense. In summer, the oceans act as refrigerators of the atmosphere. The continents (Eurasia, central and western regions of the North) are the heaters of the atmosphere in summer. America and Africa). In winter in December, on the contrary, continents have atmosphere refrigerators, oceans and ocean-continent transition zones. In the interval between winter and summer, winter and summer heat transfer structures in the atmosphere are transitional. **Figure 9** shows how the summer structure of heat flows is consistently destroyed, and how the winter structure is consistently built. Similarly, in March, the winter structure will collapse, and the summer structure is being built (the limitation of the volume of the article does not allow you to bring the corresponding maps). Here, there is a hypothesis about the existence of only two main structures of heat transfer.

At the top in **Figure 10** given the map of heat flow in an average year, identical to **Figure 6**, calculated for each mouth, and then summarized by 12 months. This is the total heat source Q. At the bottom in **Figure 10** shows a map of heat sources, as the difference between the maps of the total divergence of the heat flow Q and its average value $\overline{Q}$ for the year, that is, as the value $\tilde{Q} = (\langle Q \rangle - \overline{Q})$. The value of $\overline{Q}$ is calculated using data on overspending for the year, i.e. it is obvious that the value of $\tilde{Q}$ appears due to annual fluctuations in atmospheric temperatures and pressures. In the area of **Figure 10**, where the zones are not filled with crosses (x), seasonal fluctuations in these areas increase the constant climatic sources of heat and coolness. These positive areas coincide with the zones of powerful heat flows from the ocean to the atmosphere in the Asia-Pacific region. In general, Asia as a continent in its central and northern parts is a source of heat in the atmosphere due to seasonal temperature fluctuations and geopotential. Zones of heat flow due to seasonal fluctuations have the form of interlayers between regional and Central mainland regions (Euro-Asia).

Figure 11 shows the maps of the regions of the atmosphere at different isobaric surfaces where seasonal fluctuations in climate increase the heat sources.

In general, qualitative picture of the contribution of seasonal variations in heat sources at different heights and at ground level data in **Figure 4** is qualitatively similar to the pattern on the surface of 850 mb (**Figure 10**). Differences begin to emerge more significantly since the height of 500 mb and above.

Atmospheric zones at different heights (**Figure 11**), filled crosses, increase the release of heat from these zones due to seasonal fluctuations.

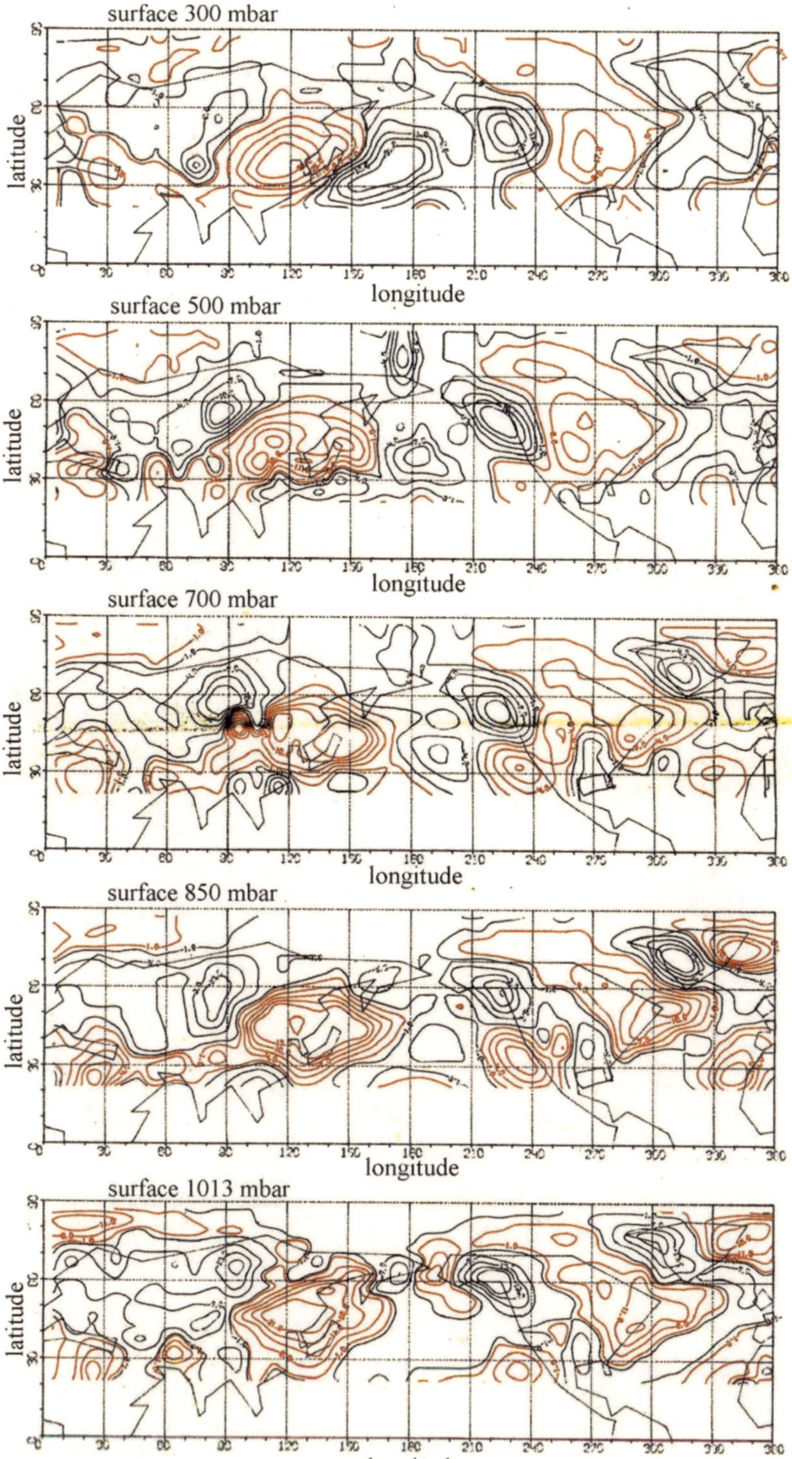

Figure 7.
Zones of positive divergence of heat flows as an area of atmospheric heating (red contour) and its cooling (black contour o on various geopotential surfaces).

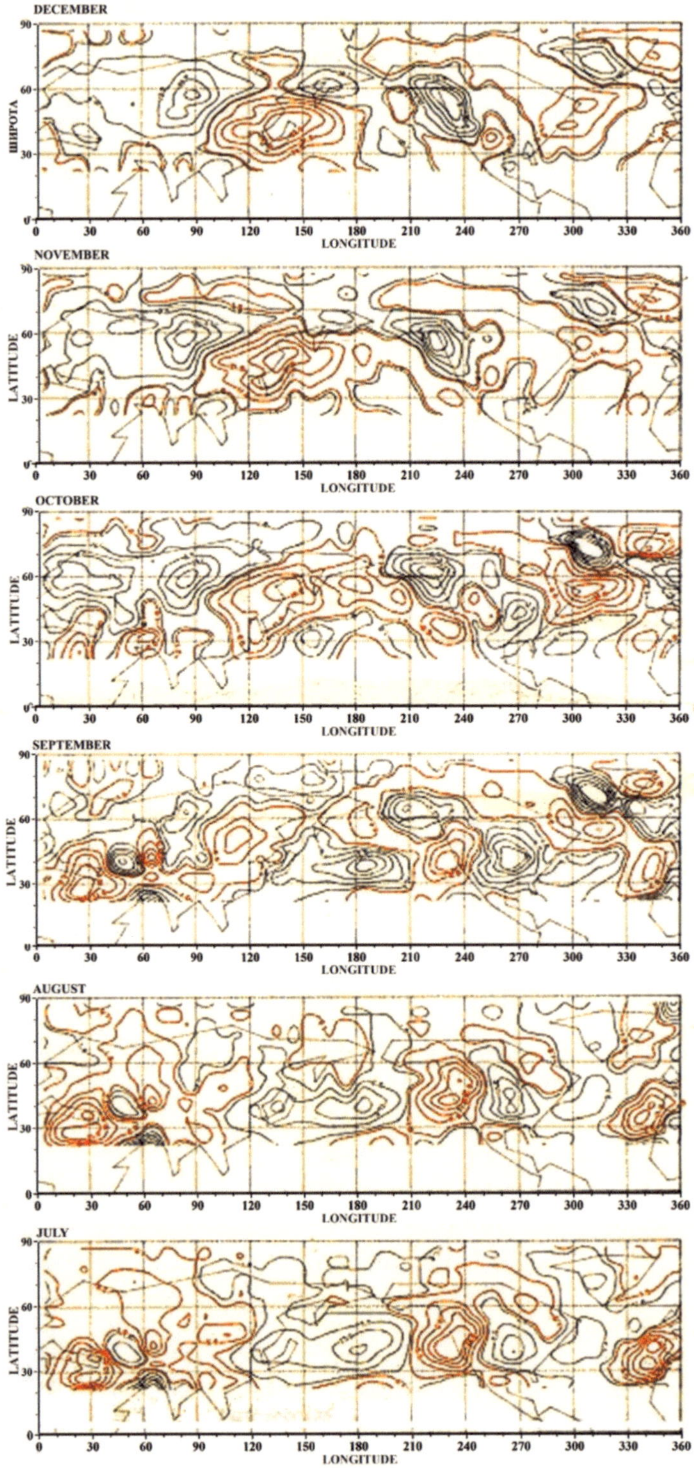

Figure 8.
Maps of heat flows on the surface of the atmosphere 850 MB in different months of the goal from December (above) to June, reflecting the evolution of the divergence structure of large-scale horizontal heat flows.

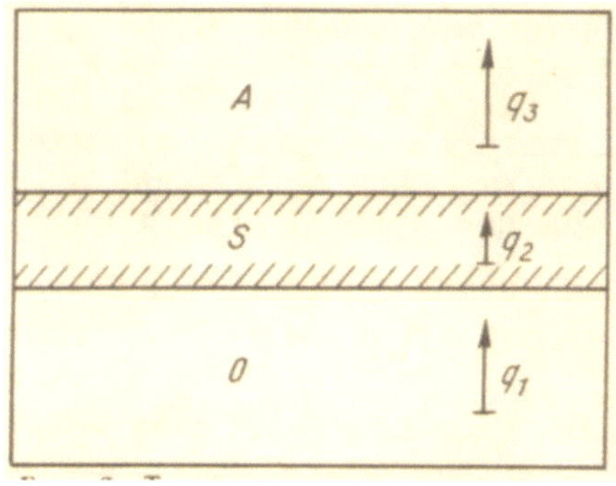

Figure 9.
Diagram of the interface of the ocean–atmosphere system. In the boundary layer of air (S), the heat flow q2 is mainly determined by the turbulent exchange at the boundary. The vertical heat flux q3 in the active layer of the atmosphere (A) is determined by convective movements. In the quasi-static conditions of the climatic seasonal course in the OSA system, the flows are equal to q1 = q2 = q3. The principle of operation of a natural heat pump pumping heat into the atmosphere is shown in the diagram.

The most powerful climatic zones removal of heat in the atmosphere (as well as the removal of "cold") are shown in the whole thickness of the atmosphere from the surface layer to altitudes of 25–30 km with a small displacement relative to each other on the heights of 1–7 km.

The geographical position of the zones of maximum and minimum divergence of the horizontal heat flow in the atmosphere experiences shifts near the climatic

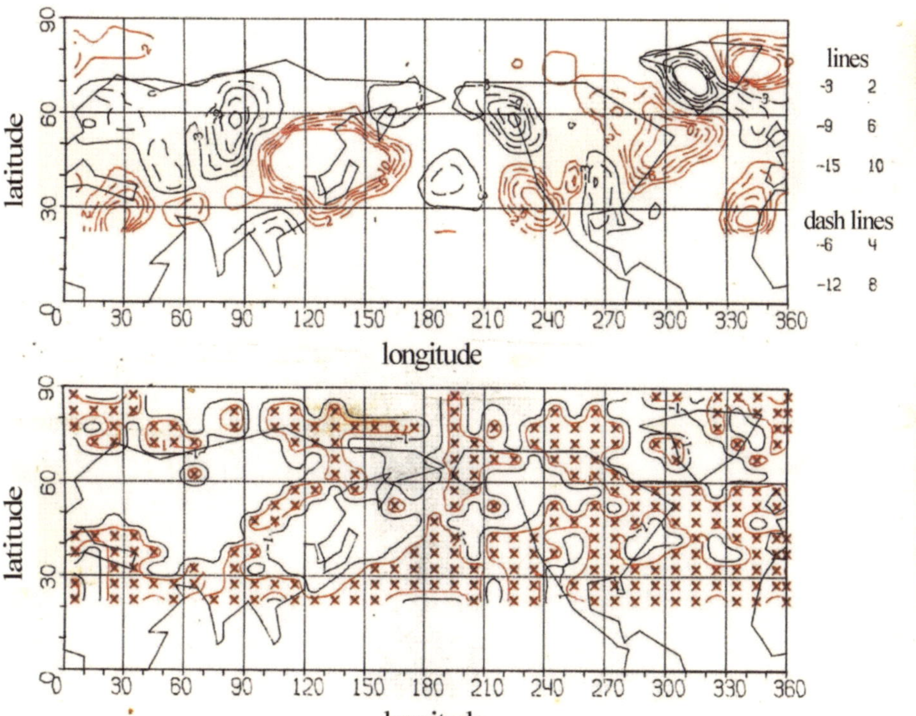

Figure 10.
The contribution of seasonal temperature fluctuations in the formation of zones of heat sources.

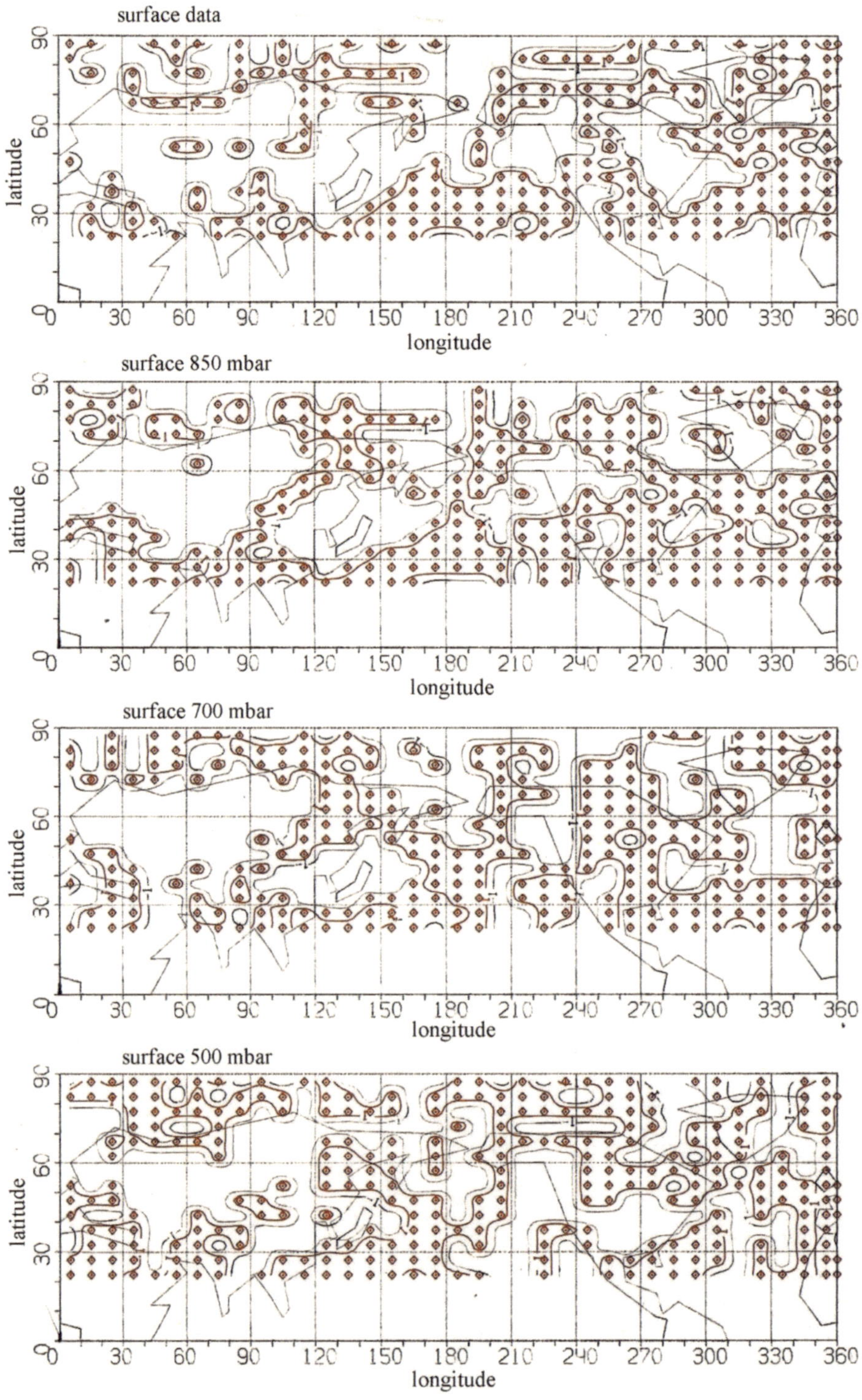

Figure 11.
The season variations, T and P, tends to increasing (or decreasing) heat zones in atmosphere.

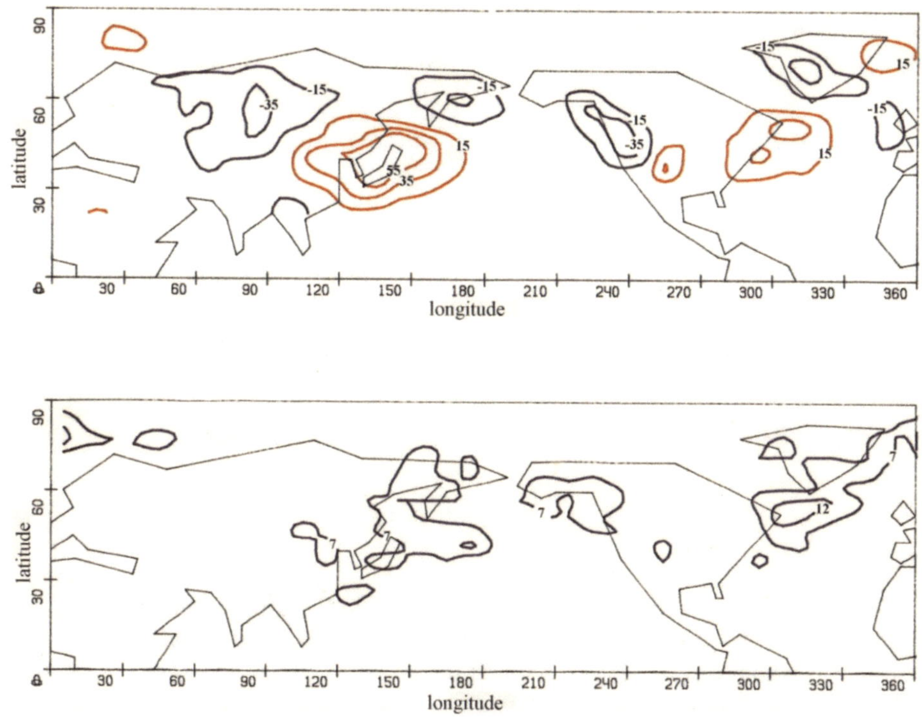

Figure 12.
Large-scale climatic (average annual) permanent zones of divergence of heat fluxes on the surface of 850 mb and the interannual variability of the Centre of the zones from year to year in the period 1975–2005. ***Figure 10*** *shows the heat flows in the January climate Bm/m², the lower one shows the areas of their shifts from year to year.*

position from year to year. The largest offsets were observed near atmospheric heating zones in the northwestern Atlantic. **Figure 10** shows the area of shifts in this area in January over a period of 30 years. Their position is "tied to the ocean–continent transition zones" (**Figure 12**).

Greenland behaves like a mainland according to the location of the zones, which can be the basis for revising the status of the island in international law.

Interannual changes in the position of the heating and cooling zones of the atmosphere do not violate their climatic structure of mutual position relative to the contour of the continents.

3. The role of air humidity in the formation of zones of divergence of heat flows

According to formulas (6) and (7), the meridional heat transfer through the contour of the Earth's latitude (L) in the geostrophic approximation has the form (7):

$$Q_m = const \oint_L \rho T V_n dL = \left(\frac{c}{\rho f}\right) \oint_L T \frac{\partial P}{\partial L} dL \qquad (7)$$

Relations (6) and (7) are applicable to the geographical contour of a large-scale atmospheric heating zone located on the ocean-continent boundary, **Figure 6**. Let us imagine atmospheric pressure as the sum of two components – dry air pressure, depending on temperature, and partial pressure of water vapor, $P = P_T + e$. Substituting this value into (7), we have (8) and (9)

$$Q_T = const\oint_L T\frac{\partial P}{\partial L}dL = const\oint_L T\frac{\partial(P_T + e)}{\partial L}dL \equiv const\oint_L T\frac{\partial(e)}{\partial L}dL \quad (8)$$

Ration (2) means that the detected large-scale zones of heat removal from the heating of the atmosphere are formed due to the mismatch of the full temperature and pressure, and the mismatch is due only to humidity. In these zones, the predominant condensation of water vapor evaporated by the oceans occurs. The removal of the flags from the atmospheric zone of removal of non-evaporated moisture (non-condensed water vapor) is associated with a mismatch of the humidity and dry airfields by the ratio (9)

$$Q_e = const\oint_L e\frac{\partial(P_T + e)}{\partial L}dL \equiv const\oint_L e\frac{\partial(P_T)}{\partial L}dL \quad (9)$$

We see a paradoxical phenomenon when in the Earth's atmosphere, with the conditional absence of moisture, the meridional large-scale heat transfer from the equator to the pole Q_{mT}=0 stops, and with the conditional disappearance of dry air pressure, the meridional transfer of water vapor Q_{me}=0 stops (8) and (9). The ratios (8) and (9) are valid for geostrophic air movements, but geostrophic movements account for about 90% of all portable atmospheric movements. The above relations (6) and (7) are valid for the calculation of large-scale heating zones in the combined cases (8) and (9).

4. Amazing efficiency of the ocean-to-atmosphere heat transfer system as a natural heat pump

Understanding this fact without the help of formulas at the level of a natural process is associated with the concept of a heat pump. In climate, it is possible to distinguish structures that can be called natural thermal machines that generate mechanical energy in the atmosphere and in the ocean due to temperature gradients as large-scale structures of geophysical convection. In the Earth's climate system, large-scale heat transfer at the water–air interface is also an analogue of a technical heat pump device as a natural heat pump (hereinafter abbreviated as NHP). A heat engine consumes thermal energy and generates mechanical energy, while a heat pump, on the contrary, consumes mechanical energy and pumps heat from a cold reservoir to a hotter one. Let us explain how the giant heat pump between the ocean and the atmosphere is manifested and functions in nature.

The scheme of heat exchange between the ocean and the atmosphere will be explained in **Figure 13**. The heat carrier in a natural heat pump is air. The heat carrier in a natural heat pump is air. The volume of air near the ocean surface is saturated with turbulent heat with ocean temperature, and has a specific lower than the overlying layers of air. The portable wind carries this volume of air higher, where it expands at a lower pressure and gives off heat to the atmosphere. Wind in a natural heat pump plays the role of its mechanical "mover". A mechanical compressor is a wind that transfers heated air from an area of high atmospheric surface pressure of 1015 mb to an area of low pressure of 850 mb, while the volume of air expands and gives heat to the atmospheric layer, which can be called the pump condensate (see **Figure 13**).

A natural heat pump (DHP) is identical in principle to a technical heat pump. In winter, it works as a heater, pumping heat from the ocean into the atmosphere, and in summer it works as an air conditioner, pumping "cold" into the atmosphere and cooling it. In this design, the atmosphere plays the role of a compressor (in the surface layer) and a coolant expander (air volume) at an altitude of about 850 mb. The portable wind is a mechanical drive and an element of the NHP operation, it provides a transfer in the atmosphere of the lifting of the driving air layer. With a wind of 5 m/sec, the volume of air rises from the surface per day to a height of about 450 meters, with a wind of 10 m/sec to a height of about 800 meters.

5. The efficiency of a natural heat pump

The efficiency (ε) of the operation of a heat pump can be determined by a simple formula $\varepsilon = \frac{Q}{W}$, where Q is the amount of heat energy pumped by the pump between the heat reservoirs per unit of time, and W is the power of mechanical energy spent on the operation of the pump. This formula is limited by a limit condition $\lim_{w\to 0} \frac{Q}{W} = \lim_{w\to 0} \frac{Q(W)}{W} \neq \infty$, the conditions for which are not related to a natural phenomenon. Let us evaluate the efficiency of a natural heat pump in the mid-latitude

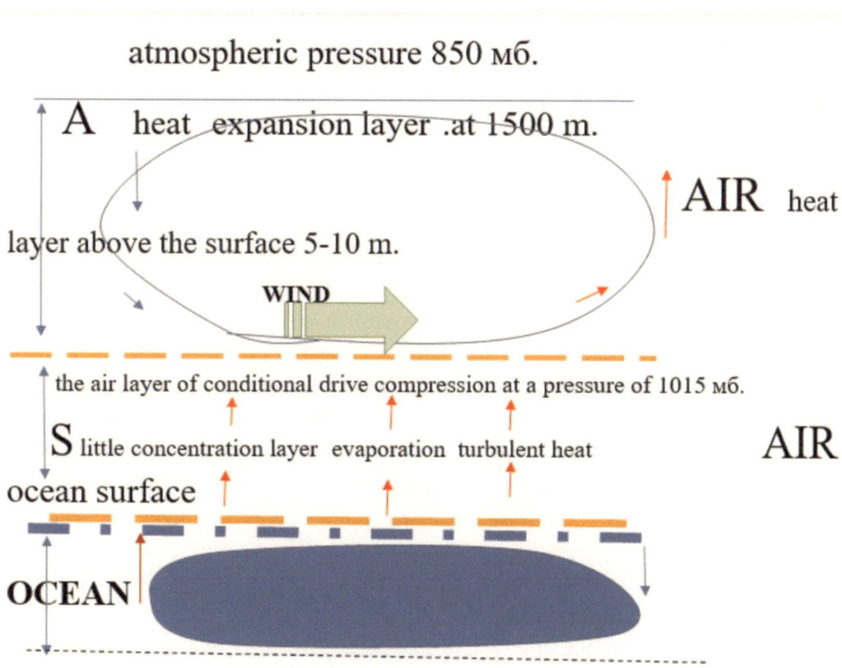

Figure 13.
The scheme of functioning of the ocean–atmosphere heat pump in nature.

zone of the northern hemisphere using this formula. The average annual heat flow in the latitudinal zone of the Atlantic 30–40 N will be taken in magnitude $Q = 50W/м^2$.

The mechanical power of the pump is equal to the every-second generation of the kinetic energy of air movement in the 1015–850 mb layer, the mass of which in this layer is M = $0.185кг/м^2$. The average speed of portable movements in this layer will be selected to the maximum value, $V = 10m/\sec$. This maximum wind speed will correspond to the minimum efficiency of the heat pump. Then the instantaneous kinetic energy of a column of air in a column with a unit surface area of the ocean (or the mechanical power of a heat pump) is equal to $W = \frac{1}{2}MV^2 = \frac{1}{2}0,185 * 100кг * м^2/\sec^2 = 9,25W/m^2$. From here, we find an estimate of the efficiency of a natural heat pump in the middle latitudes of the northern hemisphere$\varepsilon = \frac{Q}{W} = \frac{50}{9,25} \approx 5,4$. In this estimation, we assumed that all the kinetic energy of the air column movement is spent on displacement and lifting of the air volume $1м^3$ from the water–air boundary to the height of the geopotential surface 850 mb. Such an assumption is an exaggeration, therefore, the real efficiency of the phenomenon of a natural heat pump may be higher and fantastically high in some water areas, taking into account the dissipation of kinetic energy of air movement.

A heat pump consumes mechanical energy to transfer heat energy between reservoirs, including for transferring heat from a cold body to a hot one. A heat engine, on the contrary, uses thermal energy to generate reversible energy (including mechanical energy). Analogues of such machines exist in the climate as structures of natural heat engines (NHM). The efficiency of real heat engines in the framework of nonlinear thermodynamics can be calculated using the formula obtained in [9].

The formula for the efficiency of a real heat engine without any conditions for linearity or stationarity of processes has the form $\eta_{effect} = 1 - \sqrt{\frac{T_{cool}}{T_{heat}}}$, where is the temperature of the refrigerator, is the temperature of the heater. Unlike the Carnot formula and other approximations, this formula is applicable in a climate where temperatures may not be average, but instantaneous values.

For a heat engine, where the heater can be considered the air temperature at the equator $300^0к$, and in the Arctic $243^0к$, the maximum efficiency is 0.1 (10%), and with an ocean–mainland temperature difference of $10^0к$, the efficiency of the heat engine as an element of geophysical convection is wound $\eta = 1 - \sqrt{\frac{273}{283}} = 0,018$. These estimates are overestimated by an order of magnitude because they correspond to the maximum rather than average efficiency mode without taking into account losses on viscosity, turbulence and thermal losses on radiation. The real value of efficiency lies in the range $\eta \approx 0,01 - 0,001$. Thus, in the climate system, natural heat engines (NHM), with their negligible mechanical efficiency, not only soften the climate by transferring heat and smoothing temperature gradients but also activate a natural heat pump, the efficiency of which exceeds the efficiency of NHM by two orders of magnitude (or more), consuming a small part of their power.

6. Conclusions and some doubts

The heat flows from the ocean to the atmosphere, isolated as a separate component, turned out to be significant (no less than 30%). It turned out to be directly independent of the constant climatic temperatures of the ocean and atmosphere and reaches a maximum in the middle latitudes of the northern hemisphere. Therefore, it can be argued that the seasonal heat flows into the atmosphere forms and softens the climate in the middle latitudes in the same way as the climatic heat flows from the ocean.

1. The heat flows from the ocean to the atmosphere, isolated as a separate component, turned out to be significant (no less than 30%). It turned out to be directly independent of the constant climatic temperatures of the ocean and atmosphere and reaches a maximum in the middle latitudes of the northern hemisphere. Therefore, it can be argued that the seasonal heat flows into the atmosphere forms and softens the climate in the middle latitudes in the same way as the climatic heat flows from the ocean.

2. Climatic heat sources in the atmosphere (areas with an extreme divergence of heat fluxes) are not located over the continents or the oceans but are only situated at the boundary ocean-continent. For this reason, the climatic surface air temperature in the northern hemisphere is warmer than the southern, and the structure of large-scale heat transfer in the atmosphere does not coincide with the structure of the boundaries of the oceans and continents. Large-scale zone as the removal of heat in the atmosphere can be called "sources of heat" and "ventilation", and the zone's "flow" (absorption) of heat is "absorption" of heat. Large-scale atmospheric zone heat "ventilation and suction" create a regime of global "atmospheric ecology". "Ventilation" is associated with zones of removal or accumulation of impurities in the air.

3. Sources of heat in the atmosphere over the oceans are located at mid-latitude regional energy-active zones of the ocean, where the atmosphere receives heat. One part of the thermal zones of the atmosphere is part of the continent, and the other part is located in the marginal zone of the continent–ocean. The most powerful heat-divergent area atmosphere is in the East of the Asian continent, shown in **Figures 6–8** located above the ocean and over the continent.

4. The geographical structure and intensity of "thermal zones of the atmosphere" is mainly determined by the humidity (moisture content) of air.

Concept and doubts

The concept and ideas about the huge impact of seasonal-oscillatory heat transfer in the ocean–atmosphere geophysical system on the Earth's climate raise new scientific challenges and scientific doubts. The ocean–atmosphere system has a weak thermodynamic memory, and seasonal temperature fluctuations in it are not "obliged" to depend on their constant values and constant gradients. Meanwhile, such a dependence exists in the mid-latitude regions of the Earth, which indicates the mechanisms of "long-range action" between seasonal fluctuations and constant values arising in the connections of oceans, continents and the atmosphere. Such a long-range action raises both doubts and the possibility of d forecasting (scientific "foresight") short-term climate change.

Author details

Alexander Roshdestvensky
Physics-Technical Corporation LLC, Moscow, Russia

*Address all correspondence to: rojdest@rambler.ru

References

[1] Behringer D, Rieger I, Stommel H. Thermal feedback and wind stress as a of contributing cause of the gulf stream. Journal of Marine Research. 1970;**37**: 7138-7150

[2] Banker A. Contributions of surface energy flux and annual air sea interactions cycles of North Atlantic Ocean. Monthly Weather Review. 1979; **104**:9

[3] Lappo S, Gulev S, Rozhdestvensky AE. Parametrization of integral annual heat fluxes between the ocean and the atmosphere. Izvestia of the USSR Academy of Sciences, Physics of the Atmosphere and Ocean. 1985;**21**:7

[4] Lappo S, Gulev S, et al. Energy-active areas The world Ocean. DAN USSR. 1984;**275**:4

[5] Roshdestvensky A, Lappo S. Large-scale heat transfer between the ocean and the atmosphere in the annual cycle. DAN USSR. 1989;**307**(1): 88-91

[6] Roshdestvensky A. "Oscillatory" Heat Transfer and the Earth's Climate. State Oceanographic Institute. Collection of Large-scale Interaction of the Ocean and the Atmosphere and the Formation of Hydrophysical Fields. Moscow: Publishing House Hydrometeoizdat; 1989. pp. 4-18

[7] Roshdestvensky MG. Large-scale thermal zones of the atmosphere above the oceans and continents. Russian Journal of Earth Sciences. 2017;**1**:ES2001

[8] Wunsh C. The total meridional heat flux and its oceanic and atmosphere partition. Journal of Climate. 2005;**18**: 4374-4380

[9] Wallace J, Blackmon M. Observed low-frequency variability of the atmosphere. Large-scale dynamic processes in the atmosphere. M. Mir. 1998;**1998**:86-109

Chapter 5

Influence of Organic Matter on the Transport of Mineral Colloids in the River-Sea Transition Zone

Elena V. Lasareva and Aksana M. Parfenova

Abstract

The River-Sea Transition Zone has a significant impact on marine ecosystems, especially at present, due to increased anthropogenic pressure on rivers. The colloidal form of river runoff has not been practically studied, unlike the dissolved and suspended one, but this form is particularly important for the transport of river substances. The mechanisms of substance transfer were studied using model systems (colloidal clay, $Fe(OH)_3$ sol), particle aggregation was estimated by changes in optical density, turbidity and particle size. The influence of the nature of dissolved organic matter (DOM) and salinity on colloid transport was studied. It was found that humic substances (HS) (recalcitrant DOM) stabilize mineral colloids with increasing salinity, while their interaction with chitosan (labile DOM) promotes flocculation and further precipitation in the mixing zone. In natural conditions, labile DOM can be released during viral lysis of bacteria or salt stress of biota. It was shown that clay particles modified with HS are flocculated more effectively than pure clays. HS can facilitate the transport of $Fe(OH)_3$ into the outer part of the mixing zone even in the presence of flocculants. The flocculation mechanism and modern views on this process are considered.

Keywords: river-sea transition zone, colloids, clay, iron(III) hydroxide, humic substances, chitosan, extracellular polymer substances

1. Introduction

The zone of mixing of river and sea waters is an area of significant biogeochemical activity, also referred to as the river-sea transition zone (RSTZ) or marginal filter (MF) [1]. River substances enter the ocean through this region, but the possibility and extent of their input depend on the form in which these substances are present in the water in the mixing zone: dissolved, colloidal or in the form of suspended particles. The suspended component of natural water is usually defined as that which remains on the filter with pore sizes 0.45 μm after filtration and the dissolved component is the one that passes through these filters. Typically, colloid particles have a size in the range of 1 nm – 1 μm and belong to both analytically determined forms. However, according to other classifications, colloids are defined as the fraction within 1 nm–450 nm (medium-sized colloidal (200–450 nm) and fine (<200 nm) fractions [2] or from 0.2 to 1 μm isolated by ultrafiltration or field-flow fractionation techniques [3]. Due to the developed particle surface and, consequently, a large excess of surface energy, colloids tend to absorb components of river water (including pollutants) on their surface and affect their further movement into the marine environment. Despite the importance of the colloidal component it practically has not been studied, in contrast to the dissolved and suspended forms [4]. This is due to the colloid fraction being highly unstable and therefore its isolation presents certain difficulties. Especially this instability manifests itself when the salinity of the water increases.

MF can be simplified as a system consisting of three parts. In the inner part (gravitational), with a decrease in the flow rate of river water, sandy-aleuritic fractions of the suspension are deposited and coagulation of clay matter take place. At the middle part (physico-chemical) with increasing salinity to 5–12‰, the formation of aggregates begins due to flocculation of dissolved, colloidal and suspended particles. The newly formed floccules can absorb various ions, oxyhydrates and dissolved organic substances, turning them into a suspended form, as well as microalgae, bacteria and their residues [5–10]. In the outer part (biological), when the waters are significantly clarified and the concentration of biogenic elements has high values, the role of the biological component increases [11, 12]. Thus, an MF zone acts as a giant treatment plant to purify river water before it enters the sea, and the study of its mechanisms may find further application in water treatment or wastewater treatment processes.

Among the various factors determining the transport mechanisms of river substances, the role of dissolved organic matter (DOM) should be particularly noted [13–15]. Depending on the nature (recalcitrant or labile) and its concentration, DOM can act either as a flocculant that causes aggregation of mineral particles, or a stabilizer that adsorbs on the surface of particles, increases their aggregative stability and helps them move through the mixing zone [16–20]. Currently, there is an increasing interest in extracellular polymeric substances (EPS) released by marine biota that can act as flocculants, as well as in the study of the bio-flocculation process itself. Such interest was driven by the concept of green chemistry, preferring the use of non-toxic and biodegradable biopolymers as flocculants in water treatment, in particular, EPS released by marine bio-flocculant-producing bacterium [21–28]. Various bio-flocculation mechanisms are considered depending on the amount and nature of organic matter (OM), pH and ionic strength [7]. But in most of the approaches, the particles of suspension or colloids are considered in unmodified form and the nature of labile OM remains controversial.

Our approach is based on the hypothesis that recalcitrant OM (HS) of river runoff can act as stabilizers of fine mineral particles. Moreover, the HS play a dual role – they protect mineral particles in river water, and at the same time enhance bio-flocculation when meeting marine fresh OM in MF compared to unmodified particles. The key role of salinity change is not only in particle coagulation due to the compression of the double electrical layer, but also in changing biota from fluvial to marine one. We assume that a bio-flocculant in MF may be fresh OM released during bacterial viral lysis, and labile EPS compounds of marine biota under salt stress. In order to clarify the effect of individual factors on the aggregative stability of colloids, the study was carried out in artificial seawater using model systems (clay, $Fe(OH)_3$ sol). The aim of our work was to model the transport of some mineral colloid components of river runoff at increasing salinity, taking into account the effects of the nature of organic substances and to review the current state of this issue.

2. Materials and methods

Bentonite > 50% of which is represented by a fraction <1 μm (Jebel field, Turkmenistan) was used to simulate colloid clay particles.

Chitosan (Ch), natural polysaccharide, acetic form, Mw 83 kDa, deacetylation degree 85% (Bioprogress production, Shchelkovo, Russia) was used as a model flocculant (labile OM).

Humic acid (HA) potassium salt, separated from brown coal leonardite (Powerhumus, Humintech Ltd., Germany) were used as HS, modeling recalcitrant DOM.

NaCl solutions were models of water along salinity gradient in the RSTZ.

Preparation of iron hydroxide sol by Krekke method.

In a flask, 250 mL of distilled water is heated to a boil. Then, with constant stirring, 10 mL of 2% ferric chloride solution is added in small portions. The resulting sol is boiled for 2–3 min [29].

Preparation of colloidal bentonite. Colloidal bentonite (0.3 g/L) was prepared from a supernatant liquid after decantation of the settled for a week bentonite suspension (2 g/L).

Natural waters were sampled in the MF of Arctic rivers using the Rosetta Niskin bathometer, after hydro-physical parameters were determined.

Flocculation of colloid clay was carried out by the addition of Ch solution to 10 mL of clay dispersion to obtain the required concentrations.

The modification of the $Fe(OH)_3$ sol and clay colloids was made by adsorption of HA (10 mg/L) within three days. For that 0.5 mL of HA solution with concentration of 1 g/L was added to 50 mL of studied dispersions and left samples for adsorption for a week.

The size of $Fe(OH)_3$ particles was determined with a Zetatrac analyzer (Microtrac Inc., USA).

The optical density and turbidity of colloids as a measure of the aggregative stability before and after the addition of flocculant (Ch) and coagulant NaCl, were measured using the spectrophotometer Agilent (USA) at λ = 535 nm and Turbidimeter HANNA HI 98713 (Romania).

Dynamic light scattering was used to determine the particle size of colloids with Zetatrac (Microtrac Inc., USA).

3. Results and discussion

The idea of V.I. Vernadsky about boundary zones in the ocean as zones of significant biogeochemical activity is especially relevant in the mixing river-sea zone [30]. As a result of physico-chemical and biological processes, up to 93–95% of suspended and 20–40% of dissolved river matter (on average for the rivers of the world) remains in this zone [31]. However, there is not much information about the fate of the colloidal fraction.

3.1 Definition of colloidal fraction

Colloid particles transported by rivers are a mixture of mineral colloids such as metallic oxyhydroxides (mainly iron-, aluminum- and manganese oxyhydroxides), clays, carbonates and siliceous phases as well as natural and anthropogenic OM, microbiota, biogenic remains and aggregates of them with the incorporation of adsorbed cations (e.g., Ca^{2+}, Mg^{2+}, Na^{+}, etc.). The characterization of the colloidal fraction in water is difficult, in particular, because of artifacts from sampling and concentration techniques. Currently, the following methods are used to isolate the colloidal fraction of natural waters: filtration, dialysis, centrifugation, voltammetry, gels (DET/DGT), field-flow fractionation and SPLITT. Ultrafiltration membranes have been developed to separate the colloidal pool into several fractions (ranging from thousands of Daltons (Da) to hundreds of thousands of Daltons) [32]. We proposed a method for separating the colloid fraction of natural waters which has been field-tested. It was based on the coagulation of colloid particles (previously separated from the suspended matter) by electrolytes ($Al_2(SO_4)_3$ and $FeCl_3$) with the addition of $Ca(OH)_2$, transforming colloids into the particulate fraction, which was then separated by filtration through GF/F glass-fiber filters with a pore size of 0.6 μm. The addition of $Ca(OH)_2$ promoted the formation of Fe^{3+} and Al^{3+} hydroxides and, in addition, the precipitation of HS in calcium humate form [4].

3.2 Coagulation of river colloidal matter

3.2.1 Coagulation mechanism

Particles in natural waters, including OM, are negatively charged (at pH 6.0–8.0) [33]. An increase in the ionic strength of the solution with increasing water salinity in a mixing zone leads to a decrease in the surface charge of suspended and colloidal particles as a result of compression of the double electric layer and, as a consequence, their destabilization and coagulation. Due to the high surface energy of such particles, their surface in the natural environment is usually covered with a layer of adsorbed organic molecules of soil genesis, which gives them a protective function and helps to overcome the RSTZ.

3.2.2 The effect of humic substances on the coagulation of colloids

HS are the predominant group of DOM in river water, reaching ≈ 90% C_{org} with the prevalence of high molecular fraction of HS [34]. It is known that HS greatly increase the stability of bentonite colloid due to the enhancement of electrostatic repulsion and steric hindrance effects [35, 36]. Adsorption of HS may result in the stabilization of colloid system even with increasing salt concentration.

In some cases, the joint action of two or more factors enhances colloid stabilization. So, it was found that the presence of P and HA gives a synergistic effect enhancing the transfer of TiO_2 nanoparticles in columns with sand more significantly compared to that with a single presence of P or HA [37]. On the other hand, high concentration of oppositely charged ions can reduce zeta potential of colloid particles due to their adsorption and lead to destabilization of colloids containing HS. The results presented in the work [6] showed that EPS, as algal OM, enhanced the flocculation of cohesive sediments, while HS, as terrestrial OM, enhanced stabilization, rather than flocculation. This is confirmed by the observation of greater stabilization of the suspensions in the mixing zone during the period of heavy precipitation, when there is an increased intake of HS from the catchment basin. The opposite effect of EPS and HS has also been shown in this work in model tests for the ability of DOM to flocculate kaolinite suspension, which was carried out with artificial water containing purified EPS and HS.

3.2.3 Coagulation of clay colloids

The adsorption of HA on clay particles depends on the environment parameters as well as structures of clay minerals: montmorillonite (as well as bentonite which was used in our study) is characterized by a three-layered crystalline lattice, while kaolinite is characterized by a double-layer crystalline lattice and does not swell like montmorillonite. It was found that at pH 8–9, the affinity of soil HA to montmorillonite is higher compared to kaolinite, but at natural pH, the order changes and HA is better adsorbed on kaolinite [38]. In this work, it was found that the ionic strength did not affect the adsorption of HA on clay minerals with a low zeta potential. This indicates that charge screening is not the main mechanism of HA adsorption for these minerals, such as the exchange of ligands. The increase in ionic strength, however, contributed to the adsorption of HA on clay minerals with high zeta potentials. The results showed that although HA does not penetrate into the intracrystalline channels of fibrous clays, but such a structure contributes to the adsorption of HA. The study of the sorption of various fractions of HA on these clays at pH 6 and the ionic strength of NaCl 0.01 mol dm^{-3} showed that kaolinite mainly sorbs aromatic types of carbon, while montmorillonite mainly sorbs aliphatic carbon at low coverage [39]. Our earlier experiment revealed the stabilizing role of HA for both minerals without NaCl addition (4).

A laboratory study described in [18] on the effect of OM on aggregation in estuaries using montmorillonite and HA showed that river particles with a low HA content may be prone to rapid flocculation and settling in the immediate vicinity of river mouths when mixed with coastal waters rich in fresh OM. While suspended river particles rich in HA can resist immediate flocculation and be transported further away from the river mouth.

Studying the behavior of biomineral suspended particles in coastal zones with high turbidity (TMZ) and low turbidity on the shelf (OSZ), it was determined that, although the suspended particles of both zones had the same mineralogical composition, their behavior in these zones with respect to the action of bio-flocculants was different [40]. Suspended particulate matter (SPM) in TMZ formed precipitation-enriched, dense and settling biomineral aggregates, while SPM in the OSZ composed biomass-enriched, less dense, and less settleable marine snow. It can be assumed that coagulation of river material is more active in the coastal zone, while flocculation has a greater effect when particles are removed to a more marine part. Notable shifts in the molecular weight of OM from freshwater to the marine environment were noted in the work [15] in which was shown that the relatively large OM molecules in freshwater samples are transformed and/or removed effectively in the early stages of estuarine mixing.

3.2.4 Coagulation of iron(III) hydroxide sol

Fe is released during the chemical weathering of Fe-bearing minerals in the soil. During weathering under oxic conditions in the absence of organic ligands, Fe (II) is quickly oxidized to Fe (III), which in turn precipitates as oxides. Typically, the first-formed oxide mineral is amorphous, highly reactive ferrihydrite [41]. Iron oxides and hydroxides bind through various sorption reactions with the HS of the soil, forming nano-, micro- and macro aggregates that provide their protection and stabilization [42, 43]. Studying the fate of colloids along mixing zones in the Arctic, it was found that Fe exhibited a strong non-conservative coagulation/removal behavior at the beginning of the mixing zone at low salinity (<5‰), at the same time, the smallest fraction of colloid organic carbon (1 kDa-0.22 microns) remained almost constant in the entire salinity zone [44]. It can be assumed that the flocculation of iron due to newly formed organic matter is added to the coagulation of iron at the very beginning of the mixing zones.

To reveal the influence of OM on the stability of iron hydroxide sol at increasing salinity, we conducted a model experiment which showed that its aggregative stability with the addition of HS was constant up to 18‰ NaCl, after which absorbance and the particle size increased sharply from 0.510 to 2.20 and from 10 to 500 nm, respectively (**Figure 1**). The addition of HA does not lead to the destabilization of sol, although it was possible to expect such sol behavior due to the differently charged

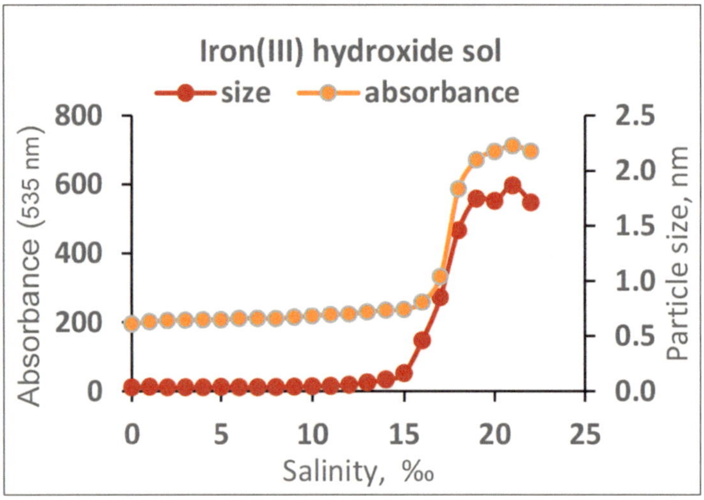

Figure 1.
Effect of salinity on the absorbance and particle size of $Fe(OH)_3$ sol.

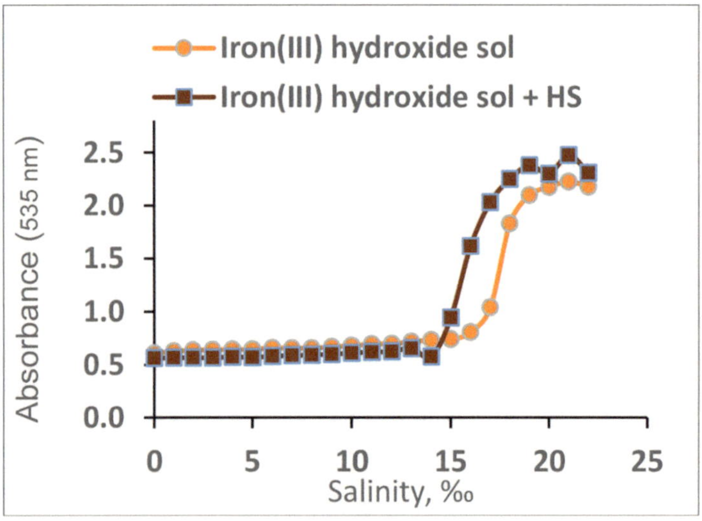

Figure 2.
Effect of salinity on the absorbance of modified with HS $Fe(OH)_3$ sol.

surfaces of sol and HA particles (**Figure 2**). The size of iron hydroxide sol particles was measured for all studied systems, and correspondence with the data on the optical density of sol was determined.

The stabilizing function of HS on a suspension of Fe(III)-montmorillonite at different pH values is shown in [45]. Comparison of steric and electrostatic effects as stabilization mechanisms showed that the steric effect is more pronounced at low pH and high ionic strength. To better understand the effect of HA on the colloidal stability of nano zero-valent iron, three types of stabilizers (polyacrylic acid (PAA), Tween-20 and starch) were used in [46]. Adsorption of HA on the surface of nano Fe particles modified with PAA enhanced the effect of electrostatic repulsion, increasing the stability of the particles. However, for nanoparticles modified by Tween-20 or starch, aggregation of particles with HA was observed. In this experiment, the effect of two organic substances on the aggregative stability of iron nanoparticles caused the different aggregation behavior and showed not only the importance of the nature of OM on the stability of particles but also the sequence of its action The study of the interaction of HA, iron oxide and bacteria in the soil showed that the interaction between iron oxides and bacteria reduces the sorption of HA [47].

3.3 Flocculation of river colloidal matter

Flocculation is the process of binding suspended particles or colloids into flocks. In the mixing zone, it occurs together with the process of coagulation. Although the flocculation potential of river water is affected by various physical (flow velocity and turbulence, temperature), chemical (pH, O_2, salinity, ions and pollutants) and biological factors (stages of aquatic biota), the key factors are the nature and concentration of the flocculants.

3.3.1 Flocculation of iron(III) hydroxide sol

To evaluate the effect of flocculation on the aggregative stability of mineral colloidal systems (clay and $Fe(OH)_3$ sol) we conducted a model study in which

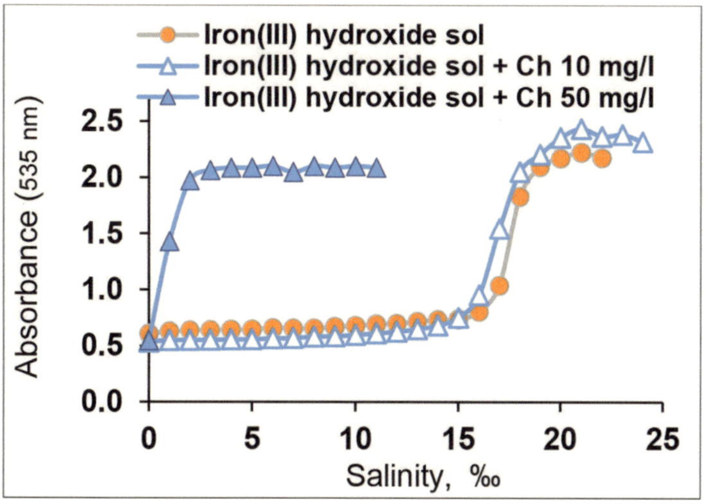

Figure 3.
Effect of salinity on the absorbance of $Fe(OH)_3$ sol with the addition of chitosan.

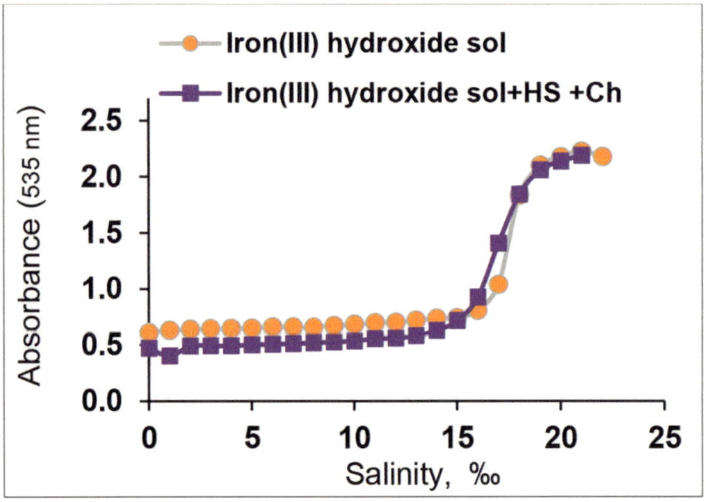

Figure 4.
Effect of salinity on the absorbance of $Fe(OH)_3$ sol modified with HS with the addition of chitosan.

chitosan (Ch) was used as a flocculant. Ch is widely used in the purification of drinking water and industrial wastewater [48, 49]. It was determined that the addition of Ch (50 mg/L) to $Fe(OH)_3$ sol leads to the formation of iron hydrosol flocks (an increase in optical density in **Figure 3**) at a salinity of 2‰, while a lower concentration of Ch (10 mg/l) did not affect the behavior of iron hydrosol. However, the behavior of the sol changes dramatically if a solution of Ch (50 mg/L) is added to $Fe(OH)_3$ sol modified with HS (**Figure 4**). In this case, the sol retains its stability up to 18‰ NaCl. HA probably creates a protective layer on iron hydrosol particles, which reduces the flocculation activity of Ch and facilitate the transport of colloidal iron to the outer part of the mixing zone even in the presence of flocculants (fresh DOM). The deposition of total Fe at 5‰, as was mentioned in [44] is probably due to the insufficient protective layer of the river iron particles, as well as the structure of HS coming from frozen soils and having rather fresh OM. This is consistent with our previous studies, in which the addition of freshwater bacteria to unmodified $Fe(OH)_3$ sol caused an increase in its particle size in the region of 5‰ from 20 nm to 500 nm due to flocculation [4].

3.3.2 Flocculation of clay colloids

Our previous tests on flocculation with clay (montmorillonite and kaolinite) showed different behavior of colloids compared with $Fe(OH)_3$ sol. The effect of Ch on modified HA clay particles led not to their stabilization but to better flocculation of colloid clay [4]. It was shown also that the increase in the hydrophobic properties of Ch results in more efficient flocculation of modified clay particles.

The turbidity of unmodified and modified bentonite colloid as a measure of flocculation over time is presented in **Figures 5** and **6**.

As follows from **Figure 6**, the flocculation of modified particles is more efficient (the gradient of the turbidity drop is greater) compared with unmodified bentonite. The results shown in **Figure 7** were obtained by repeatedly adding Ch to a colloidal solution three days later.

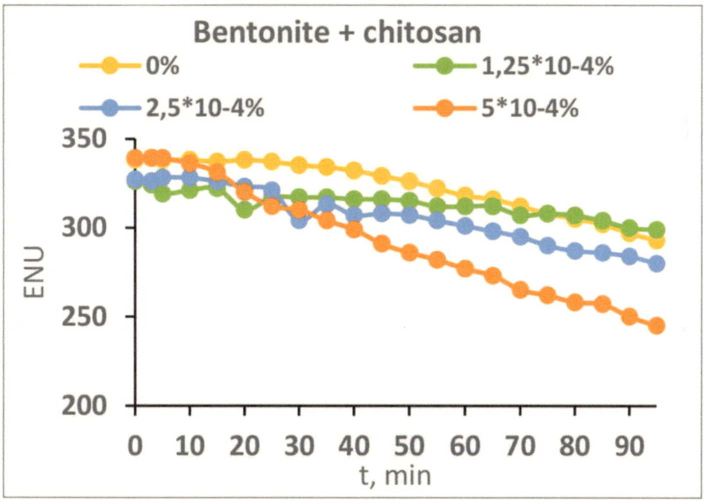

Figure 5.
Turbidity of bentonite colloid with the addition of various concentrations of chitosan as a function of time.

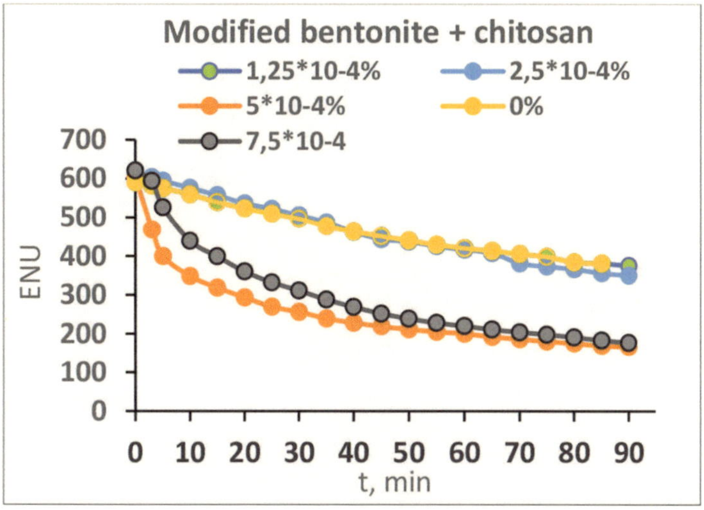

Figure 6.
Turbidity of bentonite colloid, modified with HS with the addition of various concentrations of chitosan as a function of time.

The sharp drop in turbidity after the addition of Ch 1 10–3% may indicate an increase in flocculation due to the interaction of the molecules of the newly introduced Ch with the surface-layer substances created by its early addition. The addition of NaCl leads to an even greater clarification of the solution (at Ch concentrations of $5.0x10^{-4}$% and 7.5 10^{-4}%.). Whereas Ch concentrations of $1.25x10^{-4}$% and $2.5x10^{-4}$% do not lead to flocculation of stabilized bentonite particles (**Figure 8**).

3.3.3 The composition of flocculants

Biologically mediated flocculation is currently attracting a lot of attention not only because of the study of processes in the mixing zone but also because of the search for

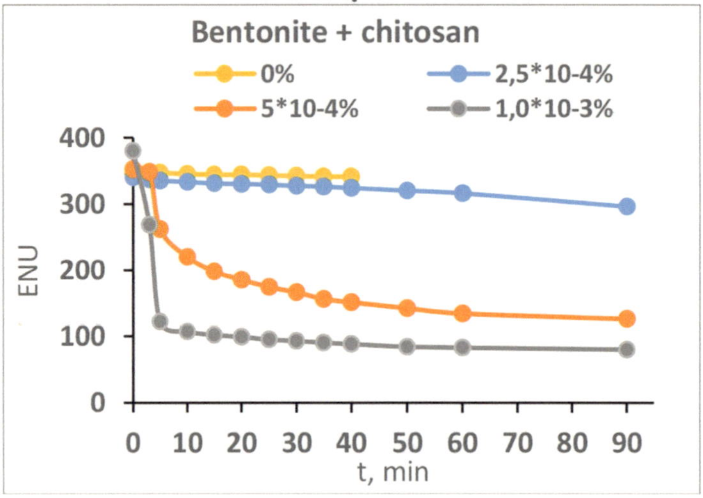

Figure 7.
Turbidity of bentonite colloid with repeat chitosan addition as a function of time.

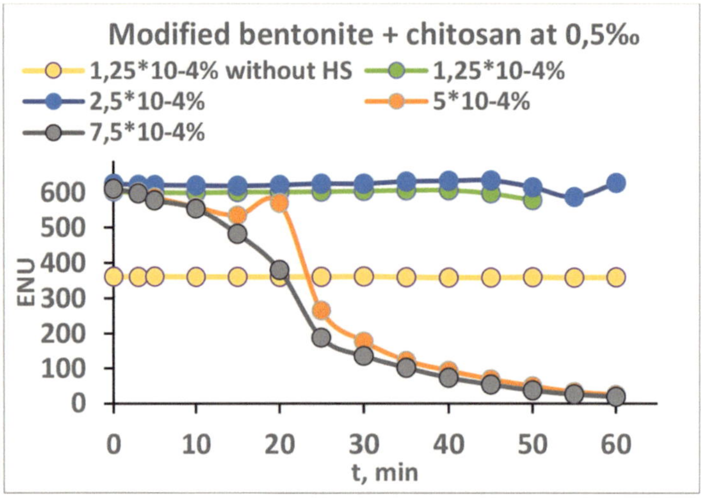

Figure 8.
Turbidity of modified with HA bentonite with chitosan addition at 0.5‰ NaCl.

environmentally safe bio-flocculants for water purification, removal of toxic organic compounds and soil remediation [7, 19, 50, 51]. EPS and/or transparent extracellular polymers (TEPs) are defined as biopolymers of microbial origin which in microbial aggregate systems are frequently responsible for binding cells and other particulate materials together (cohesion) [52]. EPS, being formed as a result of cell lysis, can be present both inside and outside microbial aggregates; usually, they consist of a mixture of macromolecules, including proteins, polysaccharides, human-like substances, nucleic acids, lipids that hold microbial cells in a three-dimensional matrix [53, 54].

As it was shown in [55], when studying EPS cyanobacterial extracellular polymeric substance, that about two-thirds of total OM within bulk EPS matrix were distributed in the HMW fraction, leaving one-third in the LMW fraction. The hydroxyl group is the most commonly mentioned functional group that promotes bio-coagulation, followed by the amine and carboxyl groups, as well as the proteins [56]. Experimental results showed the maximum flocculation activity of a natural bio-flocculant resource of *Ruditapes philippinarum* conglutination mud to kaolin clay (91.8% in seawater) that contained of 97.8% (w/w) polysaccharides and 2.2% (w/w) protein [57]. A high glucose content (47 mol. %) was noted in the DOM secreted by bacteria, while DOM, remaining after bacterial degradation, contained more galactose (33 mol. %), followed by glucose (22 mol. %) and other neutral sugars (7–11 mol. %) [58]. The chemical analysis of the purified product extracted from marine bio-flocculant the Actinomycete *Nocardiopsis aegyptia* sp. nov. revealed, that it contained mainly polysaccharides (85%) and proteins (7%) [27]. The predominantly polysaccharide composition was indicated for a bio-flocculant obtained from marine bacterium *Bacillus* sp. [59].

3.3.4 Flocculation mechanisms

Flocculation mechanisms in which biopolymers act as flocculants, such as adsorption, bridging, charge neutralization, net trapping and sweeping, as well as some other special mechanisms are summarized in [24, 49]. To clarify the role of natural OM in the aggregation of surface water particles, progressive filtration of river water samples through membranes with decreasing pore sizes (microfiltration, ultrafiltration and two degrees of nanofiltration) was used [60]. While microfiltration had a negligible effect on the spectrum of organic substances present, ultrafiltration allowed the separation of biopolymers (MW > 100 kDa), which significantly affected the size of the flakes. The authors suggest that OM with a lower Mw (<10 kDa) can create a protective layer on aluminum hydroxide particles (which was used for the coagulation test), and the modified particles are bound together by biopolymers due to a bridging mechanism. According to our view, it can be assumed that the primary adsorption layer on the surface of mineral particles in river water is formed by soil HS, which due to their amphiphilicity can be adsorbed on both hydrophobic and hydrophilic surfaces, forming a protective layer, so we modeled our system by mineral particles modified with HA (anionic polymer).

When considering a polymer mixture as flocculants, it is important to note the sequence of their addition, since pre-adsorbed polymer can have two effects: providing adsorption sites for the second polymer or blocking them, when the polycation due to the surface charge comes close to the surface of the clay particle [61]. In this case, a polyanion can have more extended adsorbed conformation. The study of the influence of cationic and anionic polyelectrolytes and their binary mixtures on the electrokinetic potential and the degree of bentonite aggregation showed that the ability to aggregate (through the formation of polymer bridges) is mainly determined by the adsorption of anionic polymer [62].

A similar result was mentioned in [63], in which a broad flocculation window was obtained due to a combination of highly charged polycation (Ch) and small amounts of high-molecular polyanion (HA), where the floccules with clay were larger compared with mono-flocculation. The results presented in [64] on the effect of various MW fractions of EPS (HW, MW and LW) obtained from wastewater on the flocculation of kaolinite and montmorillonite also confirm the effectiveness of a mixture of organic flocculants. It was determined that both the mixed EPS and the HMW-EPS demonstrate comparable flocculation characteristics in single-clay systems. However, the mixed fraction turned out to be more effective than the HMW-EPS fraction for two-component clay flocculation due to the adsorption mechanism, in which the extended anionic polymer tails of mixed EPS in solution thereby enhancing particle flocculation. The efficiency of a mixture of alginate-protamine solutions was high in flocculating kaolin and quartz particles, while neither alginate nor protamine has flocculating ability [65].

Many works devoted to flocculation in the coastal zone mention that increased flocculation is associated with the plankton bloom. To assess the influence of algae, it is proposed to introduce some indicators, such as CC/SSC ratio (the chlorophyll-a concentration to suspended sediment concentration) that increases in summer when algae-rich macro-flocs are abundant and reduced in winter when amount and size of flocks are very reduced [12]. Flocculation capability of river water was also characterized by POC/SPM ratio (where POC is particulate organic carbon). The POC/SPM value increases during the algal bloom period when stickier, particle-binding polymeric substances are formed [6]. In our study, we have proposed COC/PM and COC/CM criteria (where COC is colloid organic carbon, PM is particulate matter and CM is colloid matter) the values of which increase sharply in the flocculation zone. In combination with microbiological data, these indicators allow us to identify flocculation zones [17].

Greater effectiveness of the bio-flocculant in the presence of Ca^{2+} and Al^{3+} cations was noted in [59, 66–68]. According to the Divalent Cation Bridging Theory (DCBT), the presence of divalent cations induces bio-flocculation through bridging of negatively charged sites on EPS with cations [69]. The presence of Ca^{2+} and Mg^{2+} in seawater suggests the inclusion of these ions in the flocculation mechanism in the river-sea zone enhancing it.

The concept of biomolecular corona that formed by spontaneous adsorption of biomolecules to the highly reactive surface of nanoparticles when they enter in contact with the components of biological or environmental system was introduced to explain of colloidal stability of nanoparticles in the aquatic environment [70]. The release of EPS can also be observed during salinity stress (as cell protection), during viral lysis of microorganisms and during the death of biota species intolerant to changes in salinity [71–73]. The idea that the precipitation of river suspension in estuaries and coastal waters is significantly influenced by EPS released by microalgae is found in a number of works [13, 19, 51]. The importance of the fresh part of the TEP fraction, which contributes to the formation of larger and rapidly settling biomineral flakes in spring and summer is emphasized in [20]. However, flocculation in MF can be observed to a lesser extent in other seasons due to the reaction of biota to the change of salinity. In our opinion, the interaction of this fresh OM with mineral particles coated with recalcitrant OM (HS) leads to effective flocculation due to the synergistic effect of cationic and anionic polyelectrolytes on particle aggregation. Studies of phytoplankton in the Ob River delta have shown that the amount of DOM entering the aquatic environment as a result of viral lysis of bacteria is sufficient to trigger the flocculation process [74].

3.3.5 Field study of flocculation

Using the method described above (isolation of the colloids by electrolytic coagulation) colloidal fraction of natural waters were obtained in RSTZ of Arctic rivers (Northern Dvina, Lena and Ob) [17]. Colloids have been shown to be a key form of transport of riverine matter to the sea, two orders of magnitude higher than particulate matter. The content of the colloid component varies from 98.7 to 62.0, 96.1 to 46.4 and from 153 to 21 mg/l in the surface layer of these rivers, respectively. The waters of the Northern Dvina River are characterized by a higher content of colloidal fraction (five times higher compared to the Ob River), which is due to the nature of the water, fed from wetlands with low physical weathering The concentration of

COC decreases from the inner to the outer part of the MF with increasing salinity from 11.08 to 1.28, from 4.26 to 1.22 and from 4.65 to 1.90 mgC/L, respectively. However, at higher salinity levels, all three rivers show an increase in the colloidal fraction hundreds of times over the suspended fraction. This suggests that the main form of transfer of river material to the marine environment is the colloid fraction; the suspended fraction and the newly formed suspension remain in the MF. The low content of organic carbon in the colloidal fraction in the outer part of the MF (about 1.0–1.5%) indicates that the main form of transfer is mineral colloids.

4. Conclusions

The River-Sea Transition Zone is a natural filter system that influences the riverine colloid input to the marine environment. In this zone, due to the physico-chemical and biological processes, the substances of the river flow are transformed and most parts of the river particulate and partially dissolved matter are removed. Data on the colloidal fraction in the mixing zone are very limited due to the difficulty of isolating this fraction. It should be emphasized, however, that it is the colloidal fraction, due to its developed particle surface (reaching up to 1000 m^2/g), that has a significant effect on the behavior of other forms of riverine matter. Previously, using the method we proposed, the colloidal fraction of the water of some Arctic rivers was isolated, and it was shown that colloids significantly exceed the proportion of particulate fraction in the outer part of the MF. Among the factors influencing the transfer, one of the main ones is the content and nature of the OM. In model experiments carried out with clay colloids and iron(III) hydroxide, we have shown that, depending on the nature and concentration of OM it can lead to the stabilization or flocculation of mineral colloids. Highly transformed HS, adsorbed on clays and on iron hydroxide, stabilizes colloidal systems. However, when such coated particles meet with freshly released OM (EPS or TEP), their flocculation occurs. If Fe(III) hydroxide particles are insufficiently protected by the adsorption layer, this deposition may occur in the inner part of the missing zone at initial salinity. Otherwise, they may migrate to the outer part of the mixing zone and further towards the sea. In the case of clay colloids, the HS protective layer contributes to more efficient flocculation due to the synergistic action of polycationite (flocculant) and polyanionite (HS). The principles of natural flocculation mechanisms are already being used in the technological processes such as water and wastewater treatment, soil remediation and others ones associated with the aggregation of colloidal systems. An interdisciplinary approach to the study of the transformation of riverine colloids, and especially the study of biological processes in the mixing zone, will clarify many features of the behavior of this complex system.

Acknowledgements

The authors thank colleagues from Shirshov Institute of Oceanology of RAS for the cooperative work on the study of river-sea transition zone.

The work was carried out within the framework of the state task "Colloid chemistry" (CITIS no. 121031300084-1), and "Ecology" (CITIS no. 122040600057-3).

Author details

Elena V. Lasareva* and Aksana M. Parfenova
Department of Chemistry, Lomonosov Moscow State University, Moscow, Russia

*Address all correspondence to: elasareva@yandex.ru

References

[1] Gordeev VV, Shevchenko VV, Novigatsky VP, et al. The River–Sea transition zone (marginal filter) of the northern Dvina River as an effective trap of riverine sedimentary matter on its way to the open area of the White Sea. Oceanology. 2022;**62**(2):221-230. DOI:10.1134/S0001437022020060

[2] Fresne M, Jordan P, Daly K, Fenton O, Mellander P-E. The role of colloids and other fractions in the below-ground delivery of phosphorus from agricultural hillslopes to streams. Catena. 2022;**208**:105735. DOI:10.1016/j.catena.2021.105735

[3] Muller FLL. Exploring the potential role of terrestrially derived humic substances in the marine biogeochemistry of iron front. Earth Science Sector Marine Biogeochemistry. 2018;**6**:159. DOI: 10.3389/feart.2018.00159

[4] Lasareva EV, Parfenova AM, Demina TS, Romanova ND, et al. Transport of the colloid matter of riverine runoff through estuaries. Oceanology. 2017;**57**(4):520-529. DOI:10.1134/S0001437017040130

[5] Sholkovitz ER. The flocculation of dissolved Fe, Mn, Al, Cu, Ni, Co and Cd during estuarine mixing. Earth and Planetary Science Letters. 1978;**41**:77-86. DOI:10.1016/0012-821X(78)90043-2

[6] Lee BJ, Hur J, Toorman EA. Seasonal variation in flocculation potential of river water: Roles of the organic matter Pool. Water. 2017;**9**(5):335. DOI:10.3390/w9050335

[7] Lai H, Fang H, Huang L, He G, Reible D. A review on sediment bioflocculation: Dynamics, influencing factors and modeling. Science of the Total Environment. 2018;**642**:1184-1200. DOI:10.1016/j.scitotenv.2018.06.101

[8] Ho QN, Fettweis M, Spencer KL, Lee BJ. Flocculation with heterogeneous composition in water environments: A review. Water Research. 2022;**213**:118147. DOI:10.1016/j.watres.2022.118147

[9] Chassagne C, Safar Z, Deng Z, Heand Q, et al. Flocculation in estuaries: Modeling, laboratory and in-situ studies. In: Manning AJ, editors. Sediment Transport–Recent Advances. 2021. pp. 2-23. DOI:10.5772/intechopen.100404

[10] Khoo CLL, Sipler RE, Fudge AR. Ziegler SE. Salt induced flocculation of dissolved organic matter and iron is controlled by their concentration and ratio in boreal coastal systems, journal of geophysical research: Biogeosciences. 2022;**127**(11):e2022JG006844. DOI:10.1029/2022JG006844

[11] Deng Z, He Q, Safar Z, Chassagne C. The role of algae in fine sediment flocculation: In-situ and laboratory measurements. Marine Geology. 2019;**413**:71-84. DOI:10.1016/j.margeo.2019.02.003

[12] Deng Z, He Q, Safar Z, Chassagne C, Wang ZB. Seasonal variation of floc population influenced by the presence of algae in the Changjiang (Yangtze river) estuary. Marine Geology. 2021;**440**:106600. DOI:10.1016/j.margeo.2021. 106600

[13] Safar Z, Chassagne C, Rijnsburger S, Sanz MI, et al. Characterization and classification of estuarine suspended particles based on their inorganic/organic matter composition. Frontiers in Marine Science. 2022;**9**:896163. DOI:10.3389/fmars.2022.896163

[14] Deng Z, Huang D, He Q, Chassagne C. Review of the action of organic matter on mineral sediment flocculation. Frontiers in Earth Science. 2022;**10**:965919. DOI:10.3389/feart.2022.965919

[15] Asmala E, Massicotte P, Carstensen J. Identification of dissolved organic matter size components in freshwater and marine environments. Limnology and Oceanography. 2021;**66**(4):1381-1393. DOI:10.1002/lno.11692

[16] Blanchet M, Pringault O, Panagiotopoulos C, Lefevre D, et al. When riverine dissolved organic matter (DOM) meets labile DOM in coastal waters: Changes in bacterial community activity and composition. Aquatic Sciences. 2017;**79**:27-43. DOI:10.1007/s00027-016-0477-0

[17] Lasareva EV, Parfenova AM, Romankevich EA, Lobus NV, et al. Organic matter and mineral interactions modulate flocculation across Arctic River mixing zones. Journal of Geophysical Research: Biogeosciences. 2019;**124**(6):1651-1664. DOI:10.1029/2019JG005026

[18] Furukawa Y, Reed AH, Zhang G. Effect of organic matter on estuarine flocculation: A laboratory study using montmorillonite, humic acid, xanthan gum, guar gum and natural estuarine flocs. Geochemical Transactions. 2014;**15**:1-9

[19] Chen TY, Skoog A. Effects of salinity on abiotic aggregation of organic matter and subsequent microbial responses. Gels. 2022;**8**(12):836. DOI:10.3390/gels8120836

[20] Fettweis M, Schartau M, Desmit X, Lee BJ, et al. Organic matter composition of biomineral flocs and its influence on suspended particulate matter dynamics along a nearshore to offshore transect. Journal of Geophysical Research: Biogeosciences. 2022;**127**:e2021JG006332

[21] Selepe N, Maliehe TS, Moganedi K, Masoko P, et al. Isolation and optimisation of culture conditions for a marine Bioflocculant-producing bacterium and application of its Bioflocculant in wastewater treatment. International Journal of Environmental Research and Public Health. 2022;**19**:10237. DOI:10.3390/ijerph191610237

[22] Mu J, Cui X, Shao M, Wang Y, et al. Microbial origin of bioflocculation components within a promising natural bioflocculant resource of Ruditapes philippinarum conglutination mud from an aquaculture farm in Zhoushan, China. PLoS One. 2019;**14**(6):e0217679. DOI:10.1371/journal.pone.0217679

[23] Shahadat M, Teng TT, Rafatullah M, Shaikh Z, et al. Bacterial bioflocculants: A review of recent advances and perspectives. Chemical Engineering Journal. 2017;**328**:1139-1152. DOI:10.1016/j.cej.2017.07.105

[24] Jiang X, Li Y, Tang X, Jiang J, et al. Biopolymer-based flocculants: A review of recent technologies. Environmental Science and Pollution Research International. 2021;**28**(34):46934-46963. DOI:10.1007/s11356-021-15299-y

[25] Shukri ZNA, Cen CEC, Hossain S, Othman R, et al. A novel study on the effectiveness of bioflocculant-producing bacteria bacillus enclensis, isolated from biofloc-based system as a biodegrader in microplastic pollution. Chemosphere. 2022;**308**(Pt 2):136410. DOI:10.1016/j.chemosphere.2022.136410

[26] Siddharth T, Sridhar P, Vinila V, Tyagi RD. Environmental applications of microbial extracellular polymeric substance (EPS): A review. Journal

of Environmental Management. 2021;**287**:112307. DOI:10.1016/j.jenvman.2021.112307

[27] Mabrouk ME. Production of bioflocculant by the marine actinomycete Nocardiopsis aegyptia sp. nov. Life Science Journal. 2014;**11**:27-35

[28] Li H, Wu S, Du C, Zhong Y, et al. Preparation, performances, and mechanisms of microbial flocculants for wastewater treatment. International Journal of Environmental Research and Public Health. 2020;**17**(4):1360. DOI:10.3390/ijerph17041360

[29] Kulichikhin VG, Uchebnik V, editors. Practical Manual on Colloid Chemistry for Higher Education Students. Moscow: Vuzovskii Uchebnik; 2012. p. 282. ISBN 978-5-9558-0217-6 [in Russian]

[30] Vetrov A, Romankevich E. Disstribution and fluxes of dissolved organic carbon in the Arctic Ocean. Polar Research. 2019;**38**:350. DOI:10.33265/polar.v38.3500

[31] Lisitsyn AP. The marginal filter of the ocean. Oceanology. 1995;**34**(5):671

[32] Gaillardet J, Viers J, Dupre B. Trace elements in river waters. Treatise on Geochemistry. 2014;**7**:195-235. DOI:10.1016/B978-0-08-095975-7.00507-6

[33] Malkoske TA, Bérubé PR, Andrews RC. Coagulation/flocculation prior to low pressure membranes in drinking water treatment: A review (critical review). Environment Science Water Research Technology. 2020;**6**:2993-3023. DOI:10.1039/D0EW00461H

[34] Linnik P, Zhezherya VA, Ivanechko YS, Linnik R. Humic substances and their role in migration of metals in the high colored surface waters: The case study of rivers of the Pripyat' river basin. Russian Journal of General Chemistry. 2015;**84**(13):2572-2587. DOI:10.1134/S1070363214130143

[35] Zhen X, Niu Z, Pan D, Zhao X, et al. Mechanisms of bentonite colloid aggregation, retention, and release in saturated porous media: Role of counter ions and humic acid. Science of The Total Environment. 2021;**793**(8):148545. DOI:10.1016/j.scitotenv.2021.148545

[36] Sun Y, Pan D, Wei X, Xian D, et al. Insight into the stability and correlated transport of kaolinite colloid: Effect of pH, electrolytes and humic substances. Environmental Pollution. 2020;**266**(Pt 2):115189. DOI:/10.1016/j.envpol.2020.115189

[37] Chen M, Xu N, Christodoulatos C, Wang D. Synergistic effects of phosphorus and humic acid on the transport of anatase titanium dioxide nanoparticles in water-saturated porous media. Environmental Pollution. 2018;**243**(Pt B):1368-1375. DOI:10.1016/j.envpol.2018.09.106

[38] Chotzen RA, Polubesova T, Chefetz B, Mishael YG. Adsorption of soil-derived humic acid by seven clay minerals: A systematic study. Clays and Clay Minerals. 2016;**64**(5):628-638. DOI:10.1346/CCMN.2016.064027

[39] Elsayed MEA, Khalaf MMR, Gibson D, James RJ. Assessment of clay mineral selectivity for adsorption of aliphatic/aromatic humic acid fraction. Chemical Geology. 2019;**511**:21-27. DOI:10.1016/j.chemgeo.2019.02.034

[40] Fettweis M, Lee BJ. Spatial and seasonal variation of biomineral suspended particulate matter properties in high-turbid nearshore and low-turbid offshore zones. Water. 2017;**9**:694. DOI:10.3390/w9090694

[41] Fang Z, Wang WX. Dynamics of trace metals with different size species in the Pearl River estuary, southern China. Science of the Total Environment. 2022;**807**(Pt 1):150712. DOI:10.1016/j.scitotenv.2021.150712

[42] Iorio ED, Circelli L, Angelico R, Torrent J, et al. Environmental implications of interaction between humic substances and iron oxide nanoparticles: A review. Chemosphere. 2022;**303**(Pt 2):135172. DOI:10.1016/j.chemosphere.2022.135172

[43] Li Q, Hu W, Li L, Li Y. Interactions between organic matter and Fe oxides at soil micro-interfaces: Quantification, associations, and influencing factors. Science of The Total Environment. 2023;**855**:158710. DOI:10.1016/j.scitotenv.2022.158710

[44] Pokrovsky OS, Shirokova LS, Viers J, Gordeev VV, et al. Fate of colloids during estuarine mixing in the Arctic. Ocean Science. 2014;**10**(1):107-125. DOI:10.5194/os-10-107-2014

[45] Borgnino L. Experimental determination of the colloidal stability of Fe(III)-montmorillonite: Effects of organic matter, ionic strength and pH conditions. Colloids and Surfaces A: Physicochemical and Engineering Aspects. 2013;**423**:178-187. DOI:10.1016/j.colsurfa.2013.01.065

[46] Dong H, Lo IM. Influence of humic acid on the colloidal stability of surface-modified nano zero-valent iron. Water Research. 2013;**47**(1):419-427. DOI:10.1016/j.watres.2012.10.013

[47] Jiang L, Zhu J, Wang H, Fu Q. Sorption of humic acid on Fe oxides, bacteria, and Fe oxide-bacteria composites. Journal of Soils and Sediments. 2014;**14**(8):1378-1384. DOI:10.1007/s11368-014-0897-1

[48] Kurniawan SB, Imron MF, Chik CENCE, Owodunni AA, et al. What compound inside biocoagulants/bioflocculants is contributing the most to the coagulation and flocculation processes? Science of the Total Environment. 2022;**806**(Pt 4):150902. DOI:10.1016/j.scitotenv.2021.150902

[49] Ma'cczak. P, Kaczmarek H, Ziegler-Borowska M. Recent achievements in polymer bio-based flocculants for water treatment. Materials. 2020;**13**(18):3951. DOI:10.3390/ma13183951

[50] Huang L, Jin Y, Zhou D, Liu L, et al. A review of the role of extracellular polymeric substances (EPS) in wastewater treatment systems. International Journal of Environmental Research and Public Health. 2022;**19**(19):12191. DOI:10.3390/ijerph191912191

[51] Zhu Y, Lin M, Shen X, Fettweis M, et al. Biomineral flocculation of kaolinite and microalgae: Laboratory experiments and stochastic modeling. Journal of Geophysical Research: Oceans. 2022;**127**(11):e2022JC018591. DOI:10.1029/2022JC018591

[52] Wingender J, Neu TR, Flemming HC. What are bacterial extracellular polymeric substances? In: Wingender J, Neu TR, Flemming HC, editors. Microbial Extracellular Polymeric Substances. Berlin, Heidelberg: Springer; 1999. DOI:10.1007/978-3-642-60147-7_1

[53] Ding Z, Bourven I, Guibaud G, Hullebusch ED, et al. Role of extracellular polymeric substances (EPS) production in bioaggregation: Application to wastewater treatment. Applied Microbiology and Biotechnology. 2015;**99**(23):9883-9905. DOI:10.1007/s00253-015-6964-8

[54] Melo A, Quintelas C, Ferreira EC, Mesquita DP. The role of extracellular

polymeric substances in micropollutant removal. Frontiers in Chemistry Engineering. 2022;**4**:778469. DOI:10.3389/fceng.2022.778469

[55] Xu H, Li F, Kong M, Lv X, et al. Adsorption of cyanobacterial extracellular polymeric substance on colloidal particle: Influence of molecular weight. Science of the Total Environment. 2020;**715**:136959. DOI:10.1016/j.scitotenv.2020.136959

[56] Bouaouine O, Bourven I, Khalil F, Baudu M. Identification of functional groups of Opuntia ficus-indica involved in coagulation process after its active part extraction. Environmental Science and Pollution Research International. 2018;**25**(11):11111-11119. DOI:10.1007/s11356-018-1394-7

[57] Mu J, Zhou H, Chen Y, Yang G, Cui X. Revealing a novel natural bioflocculant resource from Ruditapes philippinarum: Effective polysaccharides and synergistic flocculation. Carbohydrate Polymers. 2018;**15**(186):17-24. DOI:10.1016/j.carbpol.2018.01.036

[58] Jørgensen L, Lechtenfeld OJ, Benner R, Middelboe M, et al. Production and transformation of dissolved neutral sugars and amino acids by bacteria in seawater. Biogeosciences. 2014;**11**:5349-5363. DOI:10.5194/bg-11-5349-2014

[59] Ntozonke N, Okaiyeto K, Okoli AS, Olaniran AO, Nwodo UU, Okoh AI. A marine bacterium, bacillus sp. isolated from the sediment samples of Algoa Bay in South Africa produces a polysaccharide-Bioflocculant. International Journal of Environmental Research and Public Health. 2017;**14**(10):1149. DOI:10.3390/ijerph14101149

[60] Su Z, Liu T, Yu W, Li X, et al. Coagulation of surface water: Observations on the significance of biopolymers. Water Research. 2017;**126**:144-152. DOI:10.1016/j.watres.2017.09.022

[61] Gregory J, Barany S. Adsorption and flocculation by polymers and polymer mixtures. Advances in Colloid and Interface Science. 2011;**169**(1):1-12. DOI:10.1016/j.cis.2011.06.004

[62] Taubaeva R, Meszaros R, Musabekov K, Barany S. Electrokinetic potential and flocculation of bentonite suspensions in solutions of surfactants, polyelectrolytes and their mixtures. Colloid Journal. 2015;**77**:91-98. DOI:10.1134/S1061933X14060179

[63] Petzold G, Geissler U, Smolka N, Schwarz S. Influence of humic acid on the flocculation of clay. Colloid and Polymer Science. 2004;**282**(7):670-676. DOI:10.1007/s00396-003-0989-8

[64] Ajao V, Fokkink R, Leermakers F, Bruning H, et al. Bioflocculants from wastewater: Insights into adsorption affinity, flocculation mechanisms and mixed particle flocculation based on biopolymer size-fractionation. Journal of Colloid and Interface Science. 2021;**581**(PtB):533-544. DOI:10.1016/j.jcis.2020.07.146

[65] Maruyama H, Seki H, Igi A. Flocculation of quartz and kaolin by alginate-protamine complex. Biochemical Engineering Journal. 2020;**162**:107713. DOI:10.1016/j.bej.2020.107713

[66] Chang M, Ma X, Dong X, Fan Y, et al. The synergistic effects of Al^{3+} and chitosan on the solid–liquid separation of coal wastewater and their mechanism of action. Polymers. 2022;**14**(19):3970. DOI:10.3390/polym14193970

[67] El-Foulani AA, Jamal-eddine J, Lekhlif B. Process Safety and

Environment Protection. 2022;**158**:400-408. DOI:10.1016/j.psep.2021.12.028

[68] Wang Q, Oshita K, Takaoka M. Harvesting Nannochloropsis oculata by chitosan and $AlCl_3$-induced flocculation: Effects of microalgal condition on flocculation performance. Bioenergy Research. 2021;**14**:924-939. DOI:10.1007/s12155-020-10201-z

[69] Higgins MJ, Novak JT. The effect of cations on the settling and dewatering of activated Sludges: Laboratory results. Water Environment Research. 1997;**69**(2):215-224. DOI:10.2175/106143097X125371

[70] Liu W, Worms IAM, Jakšić Ž, Slaveykova V. Aquatic organisms modulate the bioreactivity of engineered nanoparticles: Focus on biomolecular corona. Frontiers in Toxicology. 2022;**4**:933186. DOI:10.3389/ftox.2022.933186

[71] Steele DJ, Franklin DJ, Underwood GJ. Protection of cells from salinity stress by extracellular polymeric substances in diatom biofilms. Biofouling. 2014;**30**(8):987-998. DOI:10.1080/08927014.2014.960859

[72] Novosel N, Radić TM, Zorinc ML, Zemła J, et al. Salinity-induced chemical, mechanical, and behavioral changes in marine microalgae. Journal of Applied Phycology. 2022;**34**:1293-1309. DOI:10.1007/s10811-022-02734-x

[73] Shetty P, Gitau MM, Maróti G. Salinity stress responses and adaptation mechanisms in eukaryotic green microalgae. Cell. 2019;**8**(12):1657. DOI:10.3390/cells8121657

[74] Kopylov AI, Zabotkina EA, Sazhin AF, Romanova ND. Virioplankton in the Kara Sea: The impact of viruses on mortality of heterotrophic bacteria. Oceanology. 2015;**55**:561-572. DOI:10.1134/S0001437015040104

Chapter 6

Hydrodynamical Numerical Modeling of Coastal Areas

Joseph Harari, Renan Braga Ribeiro, Samuel Hora Yang, Matheus Souza Ruiz, Célia Regina de Gouveia Souza, Regina de Souza Ferreira and Chayonn Marinho

Abstract

Coastal areas are important, due to the great population and activities that occur there. Oceanographic conditions in coastal waters differ in many aspects of those in the open ocean: In general, spatial and temporal variations are greater. Coastal management requires environmental monitoring, commonly done by direct measurements, remote sensing, and the processing of numerical models. This chapter deals with the hydrodynamical numerical modeling of coastal areas, presenting its basic concepts, implementation and use strategies, and examples of model outputs and practical applications. These models aim to reproduce the coastal circulation, by computing the time evolution of the spatial distribution of sea level, currents, temperature, salinity, and density—considering the respective forcing of tides, winds, ocean–atmosphere exchanges and river contributions. Several examples of model outputs and coupling are presented, with maps of the distribution of temperature and salinity, which are important seawater properties; waste discharges by submarine emissaries, giving support to public health service; fate of plastic debris in the ocean, among others. Despite the enormous feasibility and utility of modern high-resolution numerical models, some warning must be considered, such as checking all the inputs and comparing the results to reliable independent information before their dissemination.

Keywords: modeling basic concepts, implementation and use strategies, model outputs and application, coastal areas, numerical modeling

1. Introduction

Coastal areas are especially important, due to the great population and the numerous activities that occur there, such as tourism, fishing, nautical sports, navigation, port activities, coastal engineering, waste discharge by submarine emissaries, among others [1]. Besides, these regions have been subject to intense pressure, associated to irregular buildings and sea level rising due to climate changes [2]. Nowadays, society does not admit polluted coastal waters, contaminated beaches, areas of marine erosion, inadequate marine buildings, or deformed landscapes. Thus, the proper management of the environmental conditions in coastal regions becomes an essential part of its economic, social, leisure, and sport activities [3–5].

Oceanographic conditions in coastal waters differ in many aspects of the conditions in the open ocean. In general, spatial and temporal variations are greater, and that is due to local factors such as river discharges, bottom topography, and coastline configuration. These effects produce greater seawater density gradients, stronger wind generated currents, higher tidal elevations, and more intense tidal currents. On the other hand, although most of the ocean dynamics is amplified in the shallow shelves when compared to the deep ocean, the energy dissipation at the shelves is much more effective, mostly due to bottom friction, i.e., friction of near bottom currents with the sea floor [6].

The direct effect of river discharge is to reduce the salinity of the surface layers and even the deep layers. In general, river discharge has significant seasonal variation, which causes seasonal fluctuations of salinity in coastal waters, much larger than in the open ocean. As rivers often carry suspended sediments, usually coastal waters are opaque; the deposition of these sediments decreases the depth, consequently resulting in navigation problems. Solar radiation penetrates to the bottom in the water columns of shallow places, causing temperature elevations higher than in the deep ocean. The wind stress acting at the ocean surface transfers more energy per volume in shallow waters. Also, the sun and moon tidal generation, being the same in the whole ocean, produce larger water transport in coastal regions. The effect of the coast as a boundary limit on ocean currents and waves is obvious, and here is one of the few situations in which man can exert significant influence on the ocean, through the construction of dikes and wave-breakers [7, 8].

Therefore, the coastal management requires qualified environmental monitoring, which is commonly done by direct measurements (including automatic equipment and gliders), remote sensing (from satellites, planes, drones, or towers), and by the processing of numerical models, both for the atmosphere and ocean. This chapter deals with the hydrodynamical numerical modeling of coastal areas, presenting its basic concepts, implementation and use strategies, and examples of model outputs and practical applications [9–12].

2. Basic concepts

The numerical models aim to reproduce the coastal circulation, by computing the time evolution of the spatial distribution of sea level, currents, temperature, salinity, and density—considering the respective forcing of tides, winds, ocean–`atmosphere exchanges, and river contributions. The models solve the basic hydro-thermodynamic equations and may be used for hindcasts and forecasts. Once the ocean dynamics is modeled, its results may be used for several other ocean models, such as pollution dispersion, waves propagation, sediments transports, and the trophic chain.

An hydrodynamical numerical model consists therefore of the solution of the basic hydrodynamic equations through numerical methods, i.e., by using computational techniques [11]. Initially, the Navier–Stokes equations consider all the physical effects that induce ocean currents, namely seawater density variations, sea surface elevation gradients, sea level atmospheric pressure gradients, wind stress at the sea surface, tidal forcing, turbulence, and internal and bottom friction. At this point, the currents are computed, through the zonal, meridional, and vertical components. Afterwards, the continuity equation computes the sea surface level; equations of heat and salt conservation give the temperature and salinity of the seawater; and finally, the seawater state equation calculates the seawater density [13].

But before solving this set of equations, the modeler should define the region of interest, which may be the whole global ocean, or a huge ocean basin (such as the South Atlantic Ocean), any shelf region (extending from the coastline to the shelf break) or even an inner estuary. Apart from the global models, the remaining ones are called limited-area models and require lateral boundary conditions. On the other hand, both global and limited-area models require the specification of surface conditions, which consist of meteorological data. In fact, coupled atmospheric–ocean models do not require specifying sea surface conditions, but these models are not the subject of this chapter.

Here, the main objective is related to limited-area models for coastal areas, from the coastline until the shelf break, excluding then deep ocean regions. And for this kind of model, the specification of lateral and surface boundary conditions is essential.

The lateral boundary conditions inform the oceanic oscillations that propagate across the open borders into the model region, consisting of the mean sea level, tidal oscillations, river discharges, and vertical profile of currents, temperature, and salinity. The sea surface conditions inform about the wind stress, the atmospheric pressure gradients, and the mass and heat exchanges at the air–sea interface [14].

Once delimited the region of interest and the sources of boundary condition data, the next step is the gridding of the region. Now the model resolution is defined, specifically the horizontal and vertical spacings for the points where the ocean parameters will be computed, once the model advances in time, with a given time step, starting from a given initial hydrodynamic condition.

Basically, two approaches are considered for gridding: Regular grids perform the computations at equally spaced points (the standard of the finite difference method), while irregular grids perform at irregularly distributed grid points (with the finite difference or the finite element methods).

The grid extension and spacing, the time step, and the period of model integration are chosen as a compromise between the available computational resources and the objectives of the numerical modeling. For example, modeling a region of 200 km x 100 km, with maximum depth of 100 m: if the user chooses an horizontal spacing of 1 km and a vertical one of 10 m, that will require much less computational effort than a choice of spacings of 100 m and 2 m, respectively. Moreover, the smaller the grid spacings, the smaller is the time step, to keep stability in the model computations. Among the objectives, they can range from very short phenomena simulations (such as a cold front intrusion during a few days) to very long ones (such as interannual variability of the ocean circulation).

After gridding the model area, bathymetric data are interpolated into the grid points, to define the wet points and levels, opposed to the dry ones (continental, island, or below the sea floor points).

Figure 1 presents two grids and respective bathymetry, used for modeling the circulation in the Shelf of Sao Paulo State (Brazil) and its inner Estuary of Santos—Sao Vicente. The former has a horizontal spacing of 1 km and the latter of 250 m. The bathymetric data were extracted from nautical charts and local hydrographic surveys. **Figure 2** presents currents computed for the large grid (1 km horizontal spacing) and for a nested grid (250 m spacing), both for the time 21 h GMT of 24/July/2021. Note that for the larger grid, one every nine computed vectors were plotted, while for the smaller grid, one every three. Evidently, high-resolution grids give much more

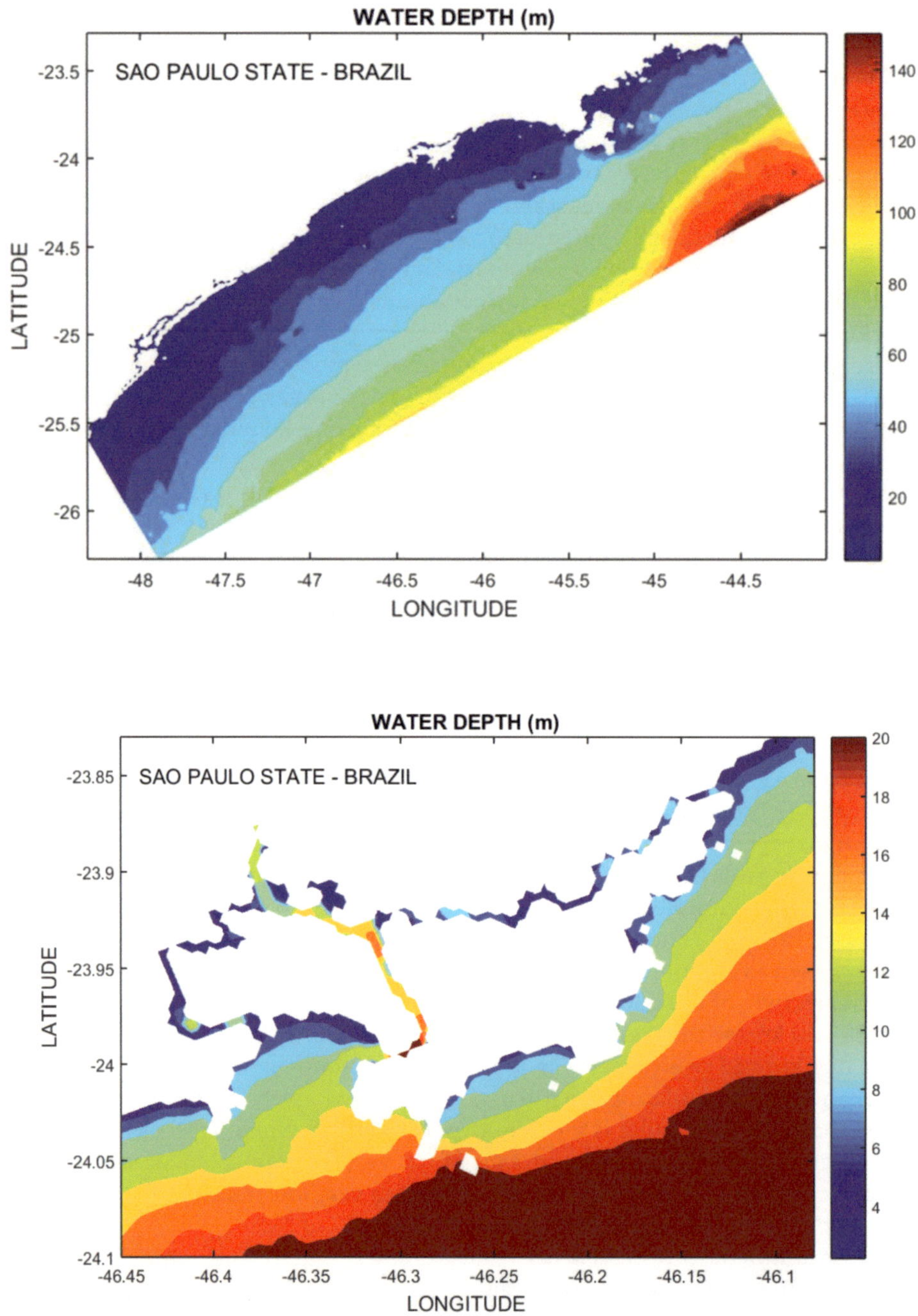

Figure 1.
Grids and respective bathymetry of Sao Paulo State Shelf (above) and Santos—Sao Vicente Estuary (below).

details of the hydrodynamics, as can be easily detected by the currents of São Vicente Channel and the Port of Santos (**Figure 2**).

At this point, it is worth to present a very useful technique used in hydrodynamic numerical modeling, named as "grid nesting". In this case, a smaller grid, with much

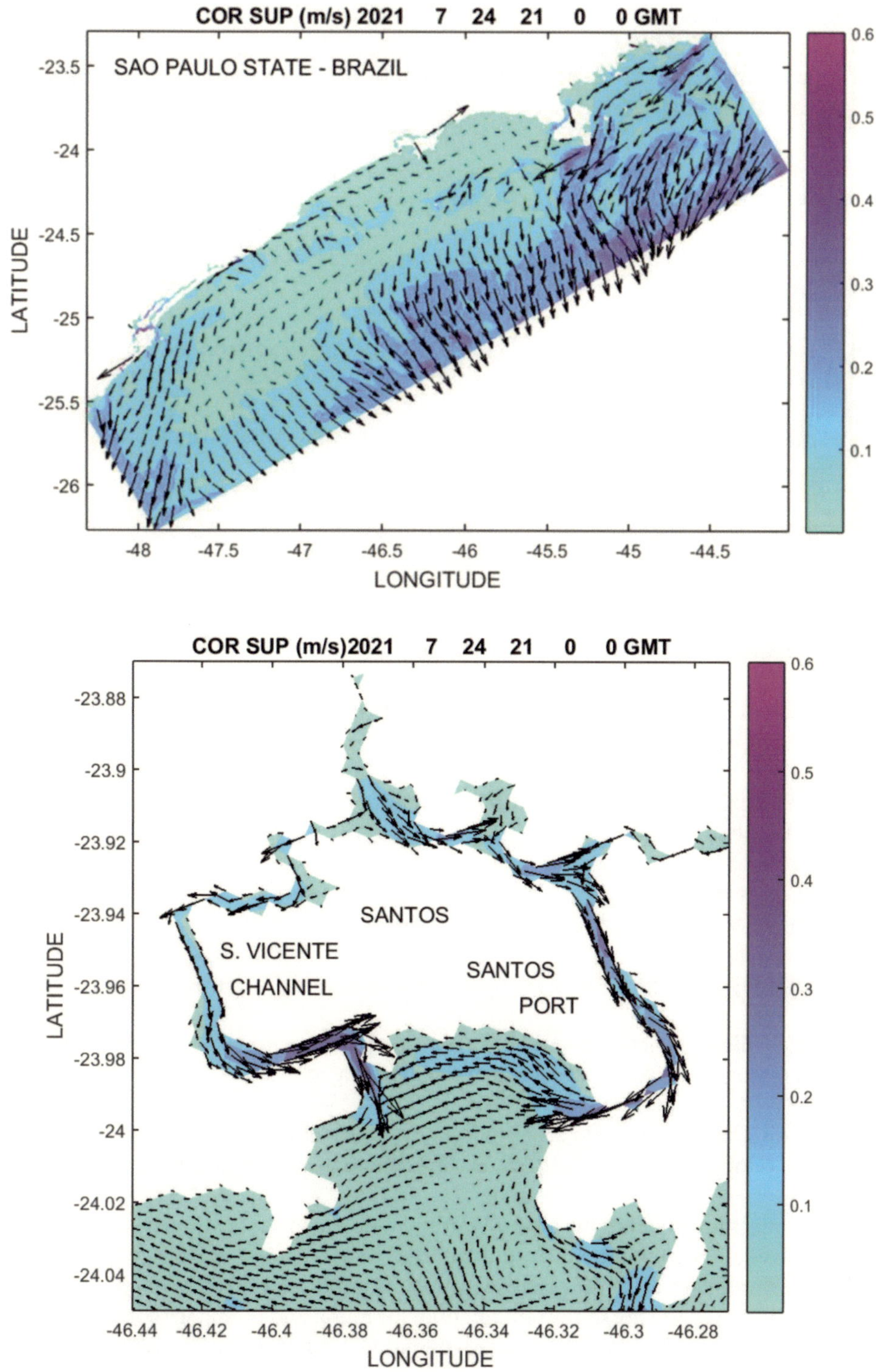

Figure 2.
Currents computed for the large grid (1 km horizontal spacing, above) and for the small grid (250 m spacing, below).

higher resolution, is nested within the bigger grid, so that the lateral boundary conditions of the smaller grid are defined by the results of the bigger one. This technique is often used when high resolution and detailed results are required only in a specific sub-region of the bigger grid.

3. Implementation and use strategies

Hydrodynamical numerical models of coastal areas may be implemented for a variety of purposes, such as:

- Forecast of sea level and coastal currents, for ocean users and navigation security [15, 16].
- Hindcast of long-term coastal circulation, to estimate coastal sediment erosion or deposition.
- Simulating of coastal currents to subsidize coastal engineering work, such as building of marinas, ports, dikes, and breakers [17].

But, most of all, the greatest utility of the hydrodynamic models is obtained when coupling them to other numerical models, such as:

- Models of surface gravity wave propagation, allowing the estimation of current–wave interactions [18].
- Water-quality models, for simulating the concentration and dispersion of pollutants in the seawater.
- Models of sediment transports, including the deposition and re-suspension of particles at the sea floor, and dredging operations [19].
- Morphodynamic models, to estimate the coastline evolution along time [20].
- Particle displacements models, for estimating the positions of missing boats, supporting boat racing, or estimating plastic trajectories.
- Trophic chain models, for studying the influence of physical parameters (currents, temperature, and salinity) in the oceanic tropic levels [21] .

Any model implementation and use strategy should consider the purpose of the modeling, the available computational resources, and the access to data. Each coupled model requires correspondent data for processing, such as pollutant spill data, or quantity of dredged materials. Data are important not only for initial and boundary conditions specification, but for calibrating and validating model results [22]. The calibration consists in comparing model results with a dataset, in order to obtain the best fitted values of the coefficients of forcing and dissipation. Once calibrated a model, validation is the comparison of its results with another dataset, to infer the quality of the model outputs.

In fact, any model, whether hydrodynamic or coupled, must have its results compared to independent information, either by direct measurements, remote sensing

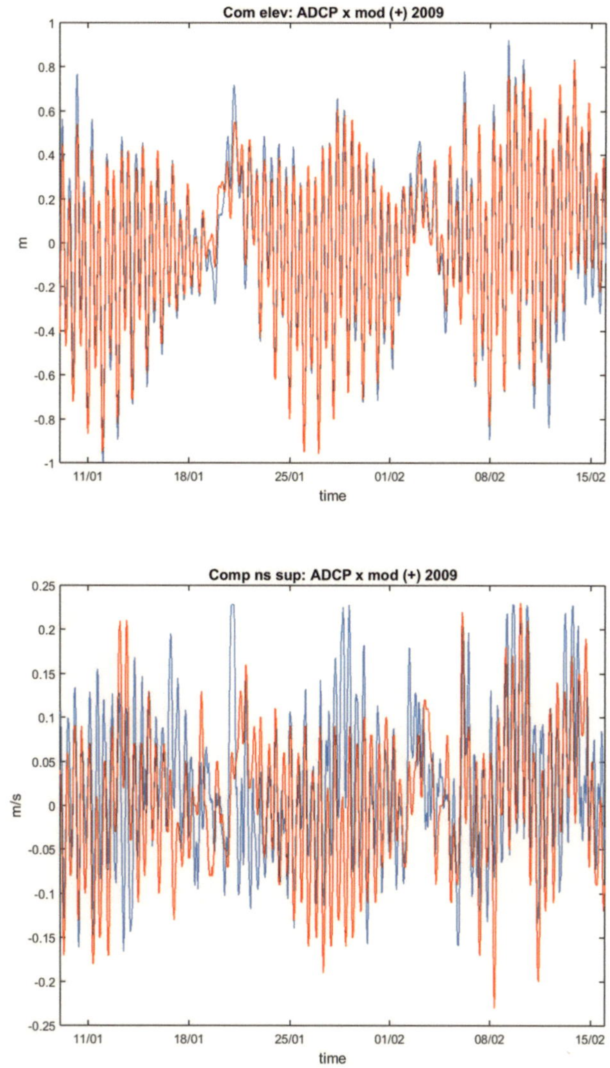

Figure 3.
Comparing model outputs (blue) to measurements (red) of sea level (above) and surface currents (below), for a simulation on January/February 2009.

data, or even other validated models. Several statistic methods are used to compare model results to independent information, such as correlation coefficient (CC) or root mean square error (RMSE) [23–25].

Figure 3 shows time series of model results of sea-level elevation and surface current component as compared to measurements, including statistical comparison parameters, at the position 24° 0,757′ S 46° 19,579′ W (coastal region of Santos). For the sea level, CC is 0.9707 ± 0.0038 and RMSE 0.0690 m; meanwhile, for the North current component the correspondent values are CC = 0.5815 ± 0.0430 and RMSE = 0.0610 m/s. This is a common feature when assessing hydrodynamic numerical model quality results: Sea-level parameters are much better than the correspondent current component ones; this is because currents exhibit much larger temporal and spatial variability than sea level.

4. Model outputs and practical applications

This section presents several examples of model outputs, with emphasis on the results of models coupled to the hydrodynamic ones (**Figures 4–6**).

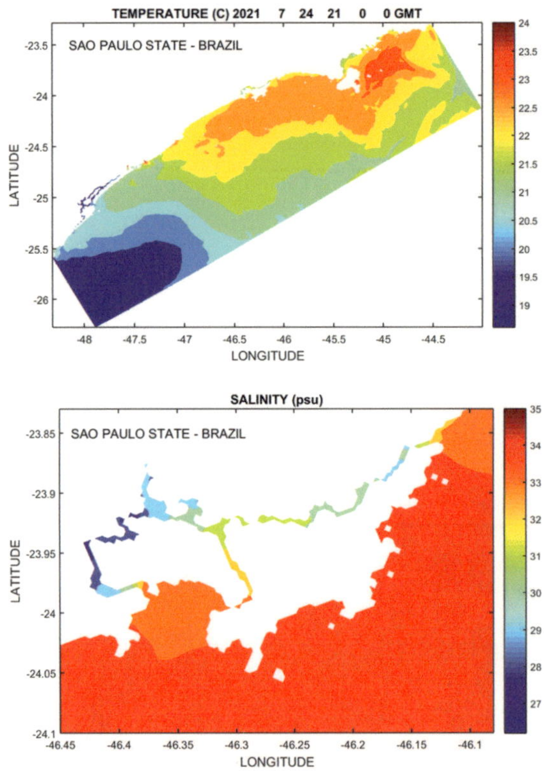

Figure 4.
Hydrodynamical numerical model computation of temperature for Sao Paulo State Shelf (above) and salinity for Santos—Sao Vicente Estuary (below), at 21 h GMT of 24/July/2021.

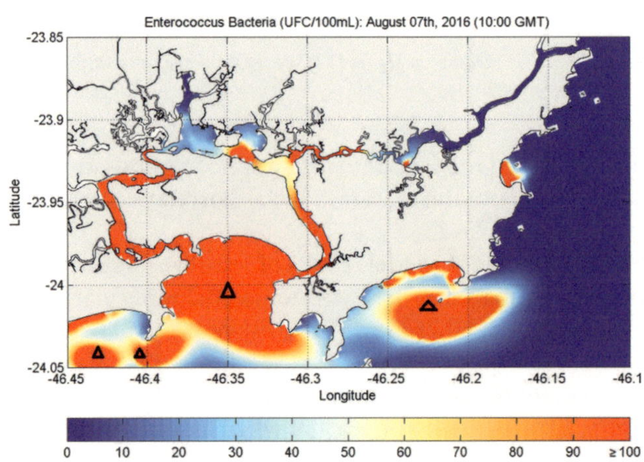

Figure 5.
Enterococus concentration as computed by a water quality model, at 10 h GMT 07/August/2016, due to the operation of submarine emissaries located at the black triangles [26].

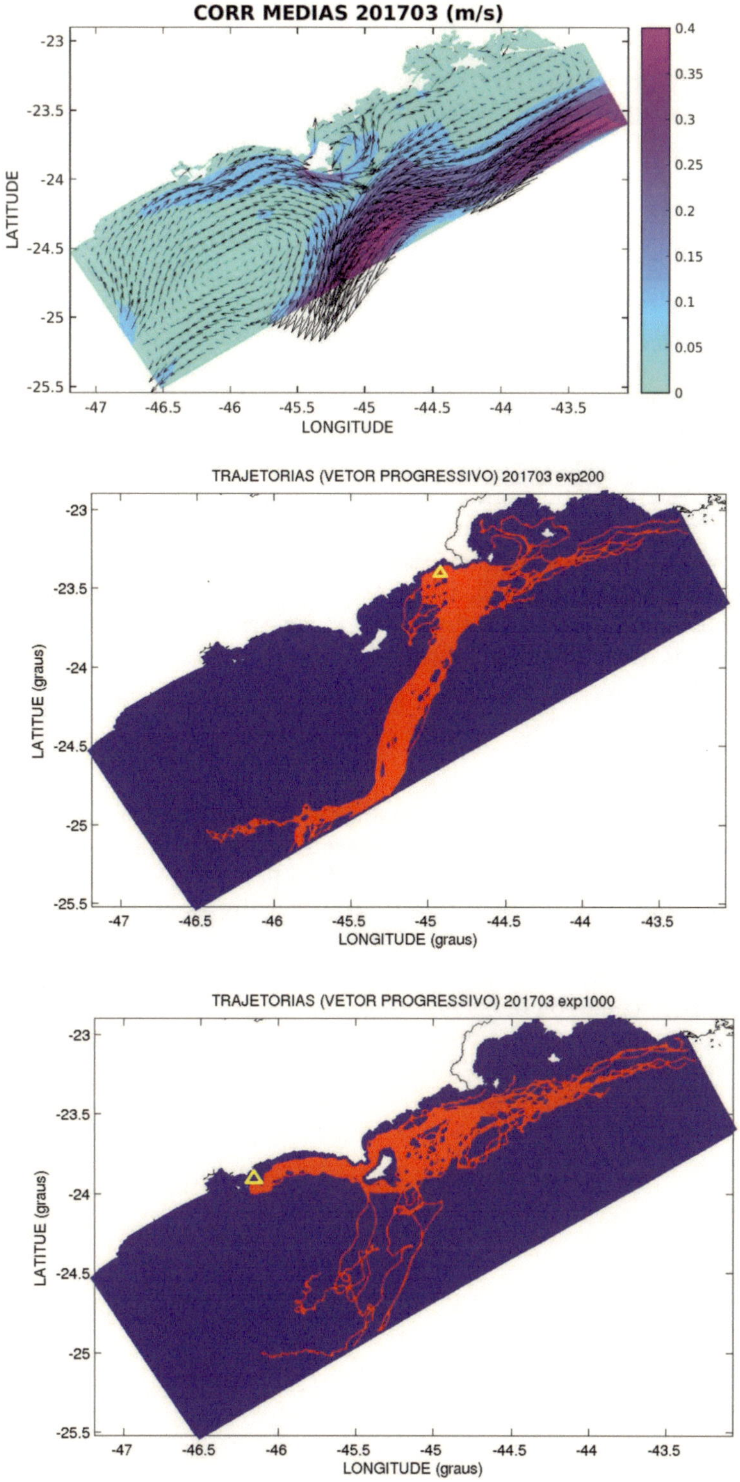

Figure 6.
Monthly mean currents at Sao Paulo State shelf in March 2017 computed by hydrodynamical model (above), and particle trajectories during this month, computed by a particle tracking model, starting at the beginning of the month on the positions marked by yellow triangles (middle and below).

5. Conclusions

Numerical modeling, and particularly hydrodynamical modeling of coastal areas, became a very powerful methodology for environmental analysis and monitoring. Several examples of model outputs and coupling were presented in this chapter.

Temperature and salinity are important seawater properties; in general, temperature has latitudinal variability, while salinity may be strongly affected by rivers, hence presenting significant variations in small distances (**Figure 4**). Monitoring of waste discharges by submarine emissaries may be performed through enterococcus concentration mapping, with real-time model processing, giving support to public health service (**Figure 5**). One of the most difficult issues of modern society is determining the fate of plastic debris in the ocean, and again numerical modeling may give an important contribution, through particle tracking models, as shown in **Figure 6**.

Despite the enormous feasibility and utility of modern high-resolution numerical models, some warning must be considered. Great care should be taken when processing a model, checking all the inputs, especially in terms of units, times, geographic positions, precision, and uncertainties. Prior to the model processing, investigating the region behavior is important, to determine the most important processes involved [27]. The model itself must be reliable, already tested by the scientific community in several different conditions. And, finally, the model results should be compared to reliable independent information before disseminating its results.

Acknowledgements

The authors acknowledge the Coordination of Improvement of Higher Education Personnel—CAPES—for the support to Project "Morphodynamic Response of Brazilian Southeast Beaches to the Effects of Sea Level Rise and Extreme Oceanographic and Meteorological Events up to 2100", process 88881.146051 / 2017-2101, and the Foundation of Support to Research of Sao Paulo State—FAPESP—for the support to Project in Public Policies "System for the Alert of Floods for the Coast of São Paulo, with a focus on Impacts of Climate Change", process 2017 / 50336-6.

Author details

Joseph Harari[1*], Renan Braga Ribeiro[2], Samuel Hora Yang[3], Matheus Souza Ruiz[2], Célia Regina de Gouveia Souza[4], Regina de Souza Ferreira[4] and Chayonn Marinho[4]

1 Oceanographic Institute of the University of Sao Paulo, Sao Paulo, SP, Brazil

2 Hydrodynamics Research Laboratory, Santa Cecília University, Santos, SP, Brazil

3 Institute of Energy and Environment of the University of Sao Paulo, SP, Brazil

4 Institute of Environmental Researches of São Paulo State, SP, Brazil

*Address all correspondence to: joharari@usp.br

References

[1] Small C, Cohen JE. Continental physiography, climate and the global distribution of human population. Current Anthropology. 2004;**45**(2):269-277

[2] IPCC – Intergovernmental Panel on Climate Change 2021. In: Delmotte V, Zhai P, Pirani A, Connors SL, Péan C, Berger S, et al, editors. Climate Change 2021: The Physical Science Basis. Contribution of Working Group I to the Sixth Assessment Report of the Intergovernmental Panel on Climate Change. Cambridge, United Kingdom and New York, NY, USA: Cambridge University Press; 2021. p. 2391

[3] Church JA, Clark PU, Cazenave A, Gregory JM, Jevrejeva S, Levermann A, et al. Sea Level Change. In: Stocker TF, Qin D, Plattner G-K, Tignor M, Allen SK, Boschung J, et al, editors. Climate Change 2013: The Physical Science Basis. Contribution of Working Group I to the Fifth Assessment Report of the Intergovernmental Panel on Climate Change. Cambridge, United Kingdom and New York, NY, USA: Cambridge University Press; 2013. DOI: 10.1017/CBO9781107415324.026

[4] Losada IJ, Gomez-Erache M, Lacambra C, Rirver E, Silva R, Toimil A. Coastal areas. In: Moreno JM, Laguna-Defior C, Barros V, Calvo Buendía E, Marengo JA, Oswald Spring U, editors. Adaptation to Climate Change Risks in Ibero-American Countries. Madrid, Spain: RIOCCADAPT Report McGraw Hill; 2020. pp. 541-580

[5] Marengo JA, Nunes LH, Souza CRG, Hosokawa EK, Pedro GR, Harari J, et al. Risk management and vulnerability to sea level rise in Brazil, with emphasis on the legacy of the METROPOLE Project in Santos. Derbyana. 2022;**43**:e768. DOI: 10.14295/derb.v43.768

[6] Simpson JH, Sharples J. Introduction to the Physical and Biological Oceanography of Shelf Seas. Cambridge: Cambridge University Press; 2012. p. 424. DOI: 10.1017/CBO9781139034098

[7] Stewart RH. Introduction to Physical Oceanography. 2008. p. 353. Available from: https://hdl.handle.net/1969.1/160216

[8] Talley LD, Pickard GL, Emery WJ, Swift JH. Descriptive Physical Oceanography: An Introduction. 6th ed. Oxford (UK): Elsevier; 2011. p. 575

[9] Kowalik Z, Murty TS. Numerical modeling of ocean dynamics. World Scientific - Advanced Series on Ocean Engineering. 1993;**5**:481

[10] Miller RN. Numerical Modeling of Ocean Circulation. Cambridge: Cambridge University Press; 2007. p. 242. DOI: 10.1017/CBO9780511618512

[11] Cushman-Roisin B, Beckers J-M. Introduction to Geophysical Fluid Dynamics - Physical and Numerical Aspects. Academic Press; 2010. p. 829. ISBN: 9780080916781

[12] Harari J. Fundamentos de Modelagem Numérica em Oceanografia. Editoração: da SALT Sea & Limno Technology; 2015. p. 246

[13] O'Brien JJ. Advanced Physical Oceanographic Numerical Modelling. Dordrecht (Netherlands): D. Reidel Publishing Company; 1985. p. 608

[14] Harari J, Camargo R. Patterns of evolution of frontal systems along the

south-southeastern Brazilian coast. In: Nunes LH, Greco R, Marengo JA, editors. Climate Change in Santos, Brazil: Projections, Impacts and Adaptation Options. Cham (Switzerland): Springer Nature; 2019. pp. 19-43

[15] Costa CGR, Leite JRB, Castro BM, Blumberg AF, Georgas N, Dottori M, et al. An operational forecasting system for physical processes in the Santos-Sao Vicente-Bertioga Estuarine System, Southeast Brazil. Ocean Dynamics. 2019;**70**:257-271

[16] Ruiz MS, Harari J, Ribeiro RB, Sampaio AFP. Numerical modelling of storm tides in the Estuarine System of Santos, São Vicente and Bertioga (SP, Brazil). Regional Studies in Marine Science. 2021;**44**:101791

[17] Dean RG, Dalrymple RA. Coastal Processes with Engineering Applications. Cambridge: Cambridge University Press; 2001

[18] Booij N, Ris RC, Holthuijsen LH. A third-generation wave model for coastal regions. 1. Model description and validation. Journal of Geophysical Research. 1999;**104**:7649-7666

[19] Zhang W. Sediment transport models. In: Harff J, Meschede M, Petersen S, Thiede J, editors. Encyclopedia of Marine Geosciences. Dordrecht: Springer; 2014. DOI: 10.1007/978-94-007-6644-0_176-1

[20] Shao D, Rappel W-J, Levine H. Computational model for cell morphodynamics. Physical Review Letters. 2010;**105**:108104

[21] Heath RM, Speirs D, Steele CJH. Understanding patterns and processes in models of trophic cascades. Ecology Letters. 2013;**17**(1):101-114. DOI: 10.1111/ele.12200

[22] Bastidas LA, Knighton J, Kline SW. Parameter sensitivity and uncertainty analysis for a storm surge and wave model. Natural Hazards and Earth System Sciences. 2016;**16**:2195

[23] Willmott CJ. On the validation of models. Physical Geography. 1981;**2**(2):184-194

[24] Willmott CJ. Some comments on the evaluation of the model performance. Bulletin of the American Meteorological Society. 1982;**63**(11):1309-1313

[25] Chai T, Draxler RR. Root mean square error (rmse) or mean absolute error (MAE)? – Arguments against Avoiding RMSE in the Literature. Geoscientific Model Development; 2014;7:1247-1250. DOI: 10.5194/gmd-7-1247-2014

[26] Yang SH. Análise de condições hidrodinâmicas e da qualidade da água na região costeira central do Estado de São Paulo através de modelagem numérica. Ph. D. Thesis – Institute of Energy and Environment of the University of São Paulo – S. Paulo, Brasil, IEE USP, supervisor: Joseph Harari; co-supervisor: Elisabete de Santis Braga, 2023, 638 p

[27] Souza CRG, Souza AP, Harari J. Long Term Analysis of Meteorological-Oceanographic Extreme Events for the Baixada Santista Region. In: Nunes L, Greco R, Marengo J, editors. Climate Change in Santos Brazil: Projections, Impacts and Adaptation Options. Cham: Springer; 2019. pp. 97-134. DOI: 10.1007/978-3-319-96535-2_6